# HANDBOOK OF AUDIO CIRCUIT DESIGN

# HANDBOOK OF AUDIO CIRCUIT DESIGN

Derek Cameron

Reston Publishing Company, Inc., Reston, Virginia
*A Prentice-Hall Company*

**Library of Congress Cataloging in Publication Data**

Cameron, Derek
    Handbook of audio circuit design.

    Includes index.
    1. Electro-acoustics—Handbooks, manuals, etc.
2. Electronic circuit design—Handbooks, manuals, etc. 3. Amplifiers, Audio—Design and construction—Handbooks, manuals, etc. I. Title.
TK5981.M47    621.38    77-16819
ISBN 0-87909-362-5

© **1978 by**
**Reston Publishing Company, Inc.**
**A Prentice-Hall Company**
**Reston, Virginia 22090**

All rights reserved. No part of this book
may be reproduced in any way, or by any means,
without permission in writing from the publisher.

10  9  8  7  6  5  4  3  2  1

Printed in the United States of America.

# CONTENTS

*Chapter 1* **Overview of Basic Design Principles**     **1**

    1-1   General Considerations, 1
    1-2   Tolerance Requirements and Calculations, 5
    1-3   Basic Device Tolerances, 14
    1-4   Principles of Power Dissipation, 16
    1-5   Law of Probability, 23
    1-6   Audio Design Pitfalls, 25
    1-7   Underwriters Laboratories (UL) Approval, 27
    1-8   Black-Box Concept, 29
    1-9   Basic Types of Distortion, 29
    1-10  Phon and Mel Units, 34
    1-11  Low-Frequency Boost, 35
    1-12  Presence Control, 37

*Chapter 2* **Fundamentals of Audio Circuit Design**     **39**

    2-1   Basic RC Circuitry, 39
    2-2   Input Impedance of an RC Section, 48
    2-3   Output Impedance of an RC Section, 49
    2-4   Filter Output Voltage, 53
    2-5   Frequency Response of Cascaded RC Sections, 55
    2-6   RC Bandpass Filter Characteristics, 57
    2-7   RC Band-Elimination (Notch) Filter, 59
    2-8   Active High/Low-Pass RC Filter Network, 61

## Chapter 3  Principles of Audio Amplifier Design          65

3-1  General Considerations, 65
3-2  Fundamental Amplifier Design Procedures, 67
3-3  Preamplification versus Power Amplification, 74
3-4  Transistor Load Lines, 78
3-5  Negative-Feedback Principles, 89
3-6  Multistage Negative-Feedback Relations, 104
3-7  Transistor Parameter Variation, 108
3-8  Production Cost Tradeoffs, 109
3-9  Notes on Musical and Speech Wave Forms, 112
3-10  Saturation Currents, 113
3-11  Bootstrapping Method of Feedback, 114
3-12  Frequency Response of Cascaded Stages, 116

## Chapter 4  Amplifier Bias Stabilization Methods          117

4-1  General Considerations, 117
4-2  Bias Stabilizing Circuits, 123
4-3  Thermistor Stabilization, 131
4-4  Diode Bias Stabilization Circuits, 135
4-5  Transistor Bias Stabilization Circuits, 141
4-6  Temperature Compensation by Breakdown (Zener) Diodes, 146

## Chapter 5  Audio Power Amplifiers          151

5-1  General Considerations, 151
5-2  Output-Transformerless Power Amplifiers, 153
5-3  Transformer Phase Inverters, 153
5-4  Quasi-Complementary Power Amplifiers, 156
5-5  Directly Coupled Load, 161
5-6  Full Complementary Output Arrangement, 166
5-7  Examples of Innovative Circuit Design, 177

## Chapter 6  Speaker Circuitry          183

6-1  Speaker Interconnections, 183
6-2  Crossover Circuitry, 185
6-3  Speaker Phasing, 190
6-4  Connections for 70.7- and 25-Volt Speaker Systems, 191

Contents                                                                    vii

*Chapter 7*   **Basic Telephone Circuitry**                                **195**
              7-1   General Considerations, 195
              7-2   Phantom Circuits, 201
              7-3   Loading Principles, 202
              7-4   Repeaters, 203
              7-5   Return Loss, 205

*Glossary*    **Acoustic, Audio-Frequency, and Sound Terms**               **207**

*Appendix 1*  **Resistor Color Code**                                      **229**

*Appendix 2*  **Capacitor Color Code**                                     **231**

*Appendix 3*  **Diode Polarity Identification**                            **233**

*Appendix 4*  **Electronic Industries Association (EIA) Preferred Values for Resistors**   **235**

*Appendix 5*  **EIA Preferred Values for Electrolytic Capacitors**         **237**

*Appendix 6*  **Power Ratios, Voltage Ratios, and Decibel Values**         **239**

*Appendix 7*  **Symbols for Field-Effect Transistors**                     **241**

*Appendix 8*  **Relation of Distortion to Positive Feedback and to Negative Feedback**   **243**

# PREFACE

The thrust of this versatile design guide is toward achieving professional competency in the engineering approach to audio circuit planning and implementation. To this end, the reader is first introduced to an overview of basic design principles. Tolerance relations and worst-case calculations are explained and illustrated, and the black-box concept is developed. In Chapter Two, fundamentals of audio circuit design are analyzed. Characteristics and tolerances on RC filter, coupling, decoupling, and bypass circuitry are described and exemplified. Principles of audio amplifier design are treated in Chapter Three. The role of negative feedback is detailed, and guidelines for production-cost tradeoffs are established.

Intensive consideration is given to amplifier bias stabilization methods in Chapter Four. Basic approaches are outlined, illustrated, and compared. Commercial practices are noted. In Chapter Five, audio power amplifier requirements are developed. Particular emphasis is placed on the essentials of output-transformerless quasi and full complementary output arrangements. Speaker circuitry is covered in Chaper Six. Various forms of crossover networks are detailed, and instructions are included for construction of standard crossover inductors. Circuitry of the 70.7- and 25-volt speaker systems used in public-address installations is provided. Basic telephone circuitry is explained in Chapter Seven, with a detailed development of hybrid coil design principles. Phantom circuits and line-loading principles are included.

It is assumed that the student has completed courses in basic electricity, basic electronics, and communication fundamentals. No mathematics other than algebra, geometry, and trigonometry is required for comprehension of the text. This book is intended to fill the gap between academic theory and basic engineering practice. Profuse illustrations have been provided to facilitate the learning process and to provide objective perspective for the student. A comprehensive glossary has been included for ready reference whenever the student is not certain of the definition for an audio technical term. The author wishes to take this opportunity to thank his fellow instructors for their constructive criticisms and suggestions. He is also indebted to the manufacturers credited throughout the text for commercial illustrative material and various technical data.

DEREK CAMERON

# HANDBOOK OF AUDIO CIRCUIT DESIGN

chapter one

# OVERVIEW OF BASIC DESIGN PRINCIPLES

## 1-1 General Considerations

Audio circuit design is primarily concerned with the planning and detailing of configurations and networks to meet particular performance specifications within the framework of reproducible design in large-scale production. Since uncompromising design techniques are generally unacceptable from the viewpoint of production costs, the audio circuit designer is confronted with the responsibility for making judicious choices of evils; that is, he is forced to accept various calculated risks. Throughout the design procedure, all component and device tolerances and ratings are evaluated within the context of maximum production *yield*. Yield is defined as the ratio of the number of usable articles at the end of a manufacturing process to the number of articles initially submitted for processing. If an article does not meet performance specifications in production test procedures, it is rejected. Basic specifications for a simple audio amplifier are exemplified in Table 1-1. Design phases are listed in Table 1-2.

**Table 1-1. Basic Specifications for a Simple Audio Amplifier**

*Response:* ±2 dB, 50–20,000 Hz.
*Inputs:* Four mikes (hi-Z, 150 kΩ; lo-Z, 200 Ω); two aux., 250 kΩ.
*Temperature range:* −20° to +70°C.
*Output:* 6 V at 2%, 4 V at 1%, 2 V at 0.5%.
*Gain:* mike (hi-Z, 60 dB), 4 mV at 4 V; (low-Z, 80 dB), 0.4 mV at 4 V output.
*Hum/noise:* −68 dB below 5 V.
*Output impedance:* 2,000 Ω.
*For 117 Vac:* Power consumption 1.2 W.

**Table 1-2. Typical Audio Design Phases**

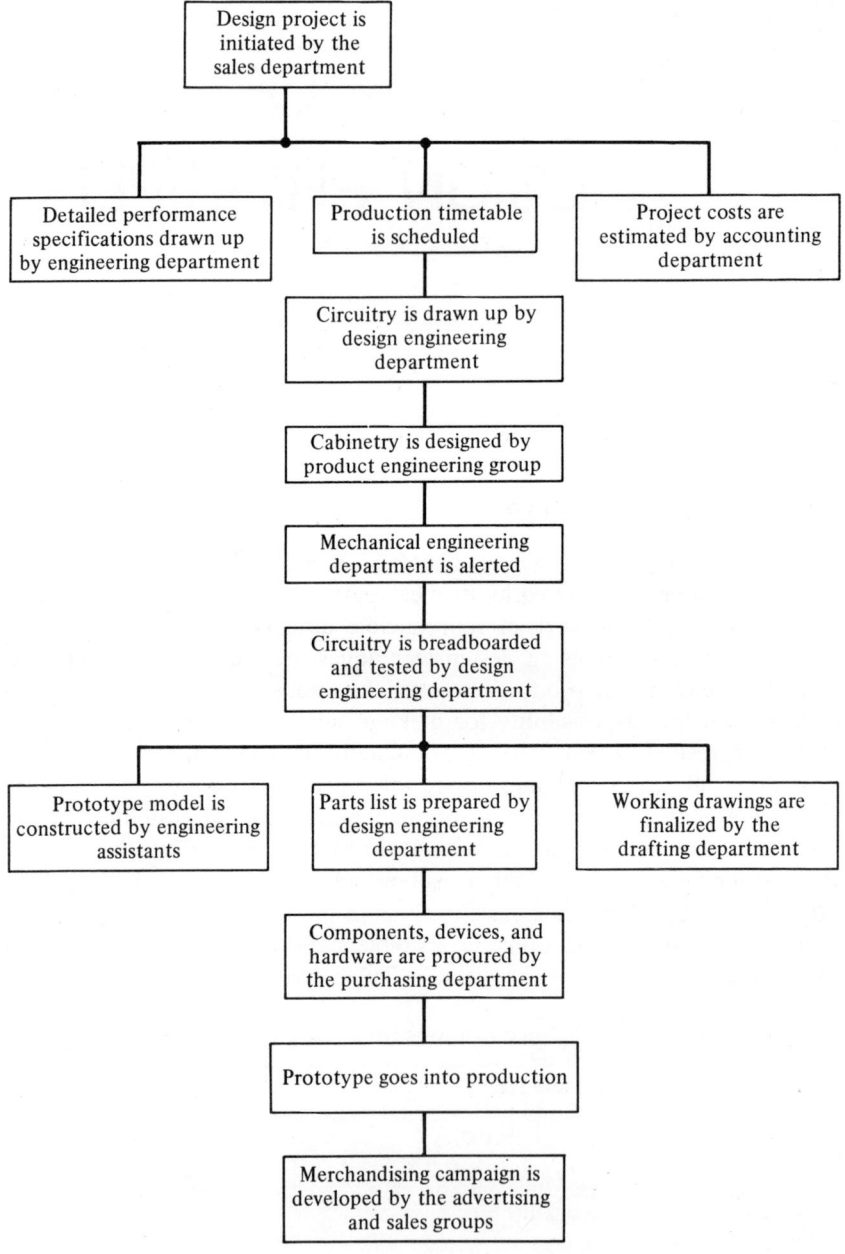

## 1-1 General Considerations

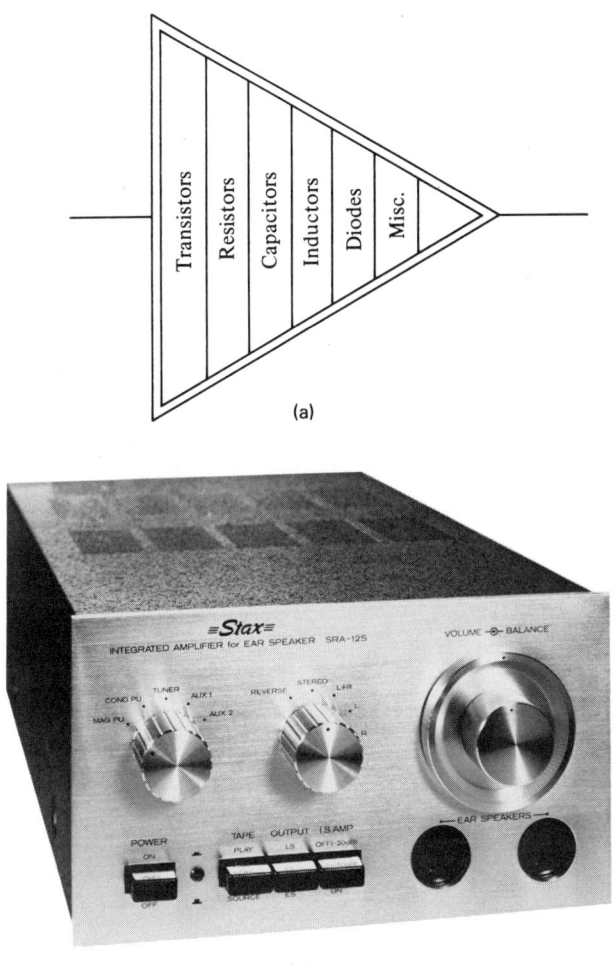

**Figure 1-1.** Basic amplifier components and devices: (a) symbolic arrangement; (b) appearance of a simple audio amplifier. (*Courtesy, Stax*)

The keystone of audio circuit design procedure is *worst-case analysis* of the *prototype model*. This is a basic design technique wherein circuit performance is determined for the case in which all components and devices (Fig. 1-1) have simultaneously assumed their most unfavorable tolerance values, in which the supply voltage is minimum, and in which environmental conditions (temperature, for example) are most adverse. A prototype model is a working model, usually hand assembled, that is suitable for complete evaluation of mechanical and electrical form, design, and performance. Approved parts are employed throughout so that it will be completely representative of the final, mass-produced equipment. A simplified representation of worst-case analysis is shown in Fig. 1-2. In the case of a complex audio network, it is sometimes difficult to calculate the particular combination of positive and negative tolerances that will produce the maximum detrimental changes in the output.

A *tolerance* is a permissible deviation from a specified (bogie) value. In other words, a *bogie value* is a design-center value. Thus resistors may be rated for a tolerance of ±20 percent, or ±10 percent, or ±1 percent. Close-tolerance (tight-tolerance) resistors are more expensive than those with relaxed tolerances. Semiconductor devices ordinarily have wide tolerances; however, the circuit designer may specify tight-tolerance devices. Or he may specify *matched pairs* of devices; in this case, a given pair of devices will have closely similar characteristics, although their bogie value will not be the same as that of another given pair of devices. Matched pairs of devices are used in push–pull audio output stages, for example. Note that resistive tolerances may be of little concern in one type of circuit,

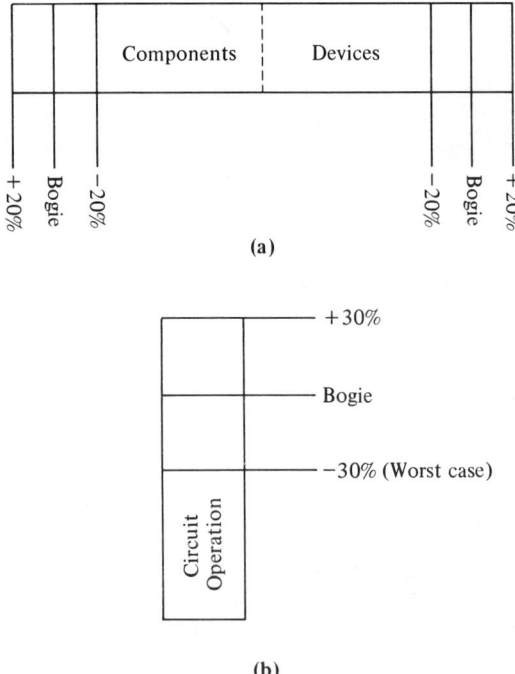

**Figure 1-2.** Hypothetical example of worst-case analysis: (a) components and devices have individual tolerances of ±20 percent; (b) circuit operation deteriorates by 30 percent in worst case.

but of great importance in another. These considerations can usually be evaluated, or at least approximated, from variational analysis of the design equations.

Most electrical tolerances have a finite life expectancy; in turn, the audio circuit designer also evaluates *reliability* requirements, and allows a corresponding margin for component and device deterioration. Reliability involves the theory of probability; it is defined as the probability that a device will perform adequately for the length of time intended under the operating environment encountered. *Quality-control* procedures must often be employed to avoid the hazard of ruinous deviations or faults in production runs. Quality control is defined as the control of variation in workmanship, processes, and materials in order to produce a consistent, uniform product. Incoming inspecting procedures may be established in various situations to ensure that components, devices, and hardware are within specified tolerances before they go into production.

## 1-2 Tolerance Requirements and Calculations

Components and devices have exact values and ideal characteristics only in theory. In practice, every fabricated unit is subject to tolerances. For example, a composition resistor may have a rated resistance value of 10,000 ohms ($\Omega$), with a tolerance of $\pm 10$ percent. This tolerance rating denotes that the actual value of a resistor chosen at random in a production lot could range in value from 9,000 to 11,000 $\Omega$. Consider the significance of this tolerance on the 266-kilohm (k$\Omega$) bias resistor in Fig. 1-3. The emitter–base junction is biased at 0.2 volt (V) and 33 microamperes ($\mu$A). Since the base–emitter voltage (0.2 V) is approximately 2 percent of the bias-resistor voltage (8.8 V), the following conclusions can be drawn in a first analysis:

1. The bias current is determined chiefly by the value of the bias resistor.
2. If the bias resistor has a tolerance of $\pm 10$ percent, the resulting tolerance on the bias current will be virtually $\pm 10$ percent.
3. If the base–emitter junction resistance has a tolerance of $\pm 10$ percent, the resulting tolerance on the bias current will be practically zero.

In other words, the bias current in this configuration is obtained from an approximate *constant-current* (current) source. Note that an ideal constant-current source would have to employ a bias resistor with infinite resistance. The basic distinction between a *constant-voltage* (voltage)

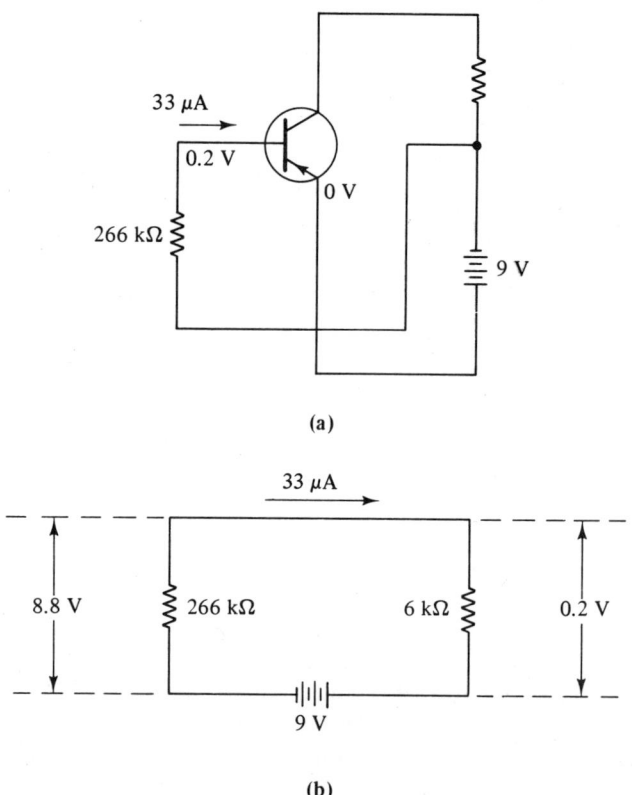

**Figure 1-3.** Elementary transistor bias arrangement: (a) configuration; (b) equivalent circuit.

source and a constant-current (current) source is depicted in Fig. 1-4. *In a constant-voltage arrangement, the load voltage is practically independent of the load-resistance value. On the other hand, in a constant-current arrangement, the load current is practically independent of the load-resistance value.* Thus tolerance considerations in circuits with voltage sources are not the same as in circuits with current sources.

Resistors with equal tolerance values combine in configurations that have the same tolerance value on their net or total resistance, as exemplified in Fig. 1-5. Thus, in Fig. 1-5a, the net resistance of a series combination of 20 percent resistors has a tolerance of 20 percent. Next, in Fig. 1-5b, the net resistance of a parallel combination of 20 percent resistors has a tolerance of 20 percent. It follows that the net resistance of a series–parallel combination of 20 percent resistors, such as depicted in Fig. 1-5c, will have a tolerance of 20 percent. In generalized form, shown in Fig. 1-5d, the net resistance of a resistive T section with a load resistor $R_L$, in which each resistor has a 10 percent tolerance, will have a tolerance of 10 percent.

## 1-2 Tolerance Requirements and Calculations

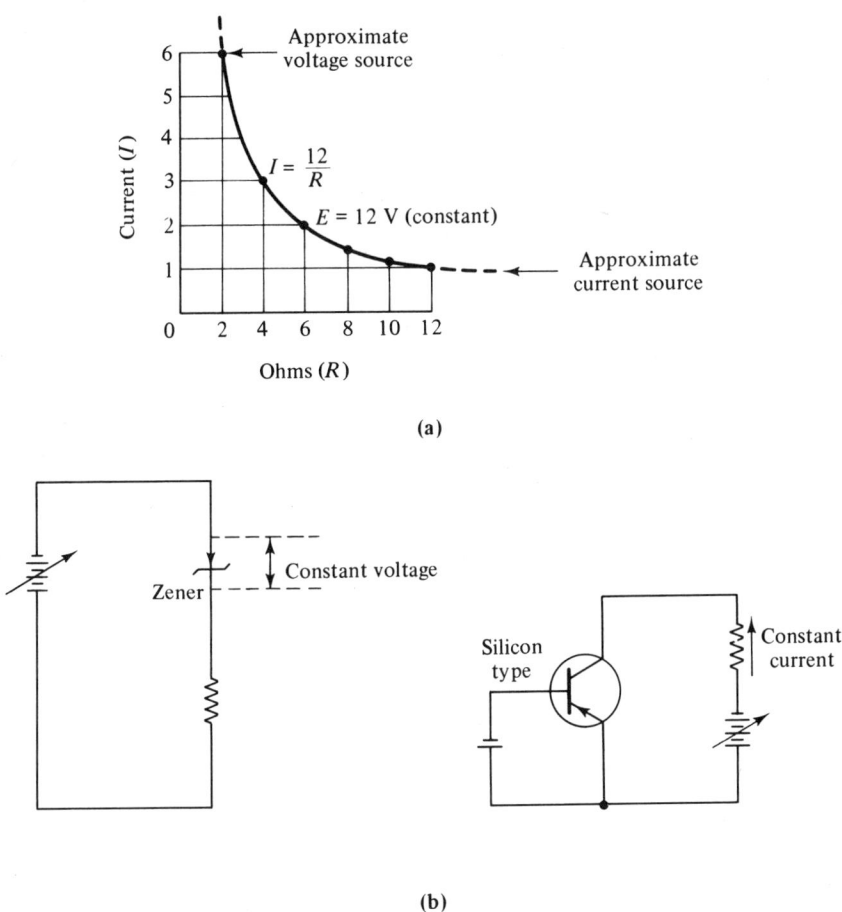

**Figure 1-4.** Basic distinction between a practical constant-voltage source and a practical constant-current source: (a) characteristic of component configuration; (b) semiconductor constant-voltage and constant-current sources.

Inasmuch as any resistive configuration whatsoever can be reduced to an equivalent T section (insofar as input–output relations are concerned), it is evident that resistors with the same tolerance value will combine in any conceivable network to present a net resistance that has the same tolerance as that of its component resistors.

Observe that, if the resistors in a series or parallel configuration have unequal tolerances, their net resistance will have a tolerance value that is intermediate to the extreme tolerance limits of the component resistors. As an illustration, if a 10-k$\Omega$ ± 10 percent resistor is connected in series with a 10-k$\Omega$ ± 20 percent resistor, their net resistance will have a value

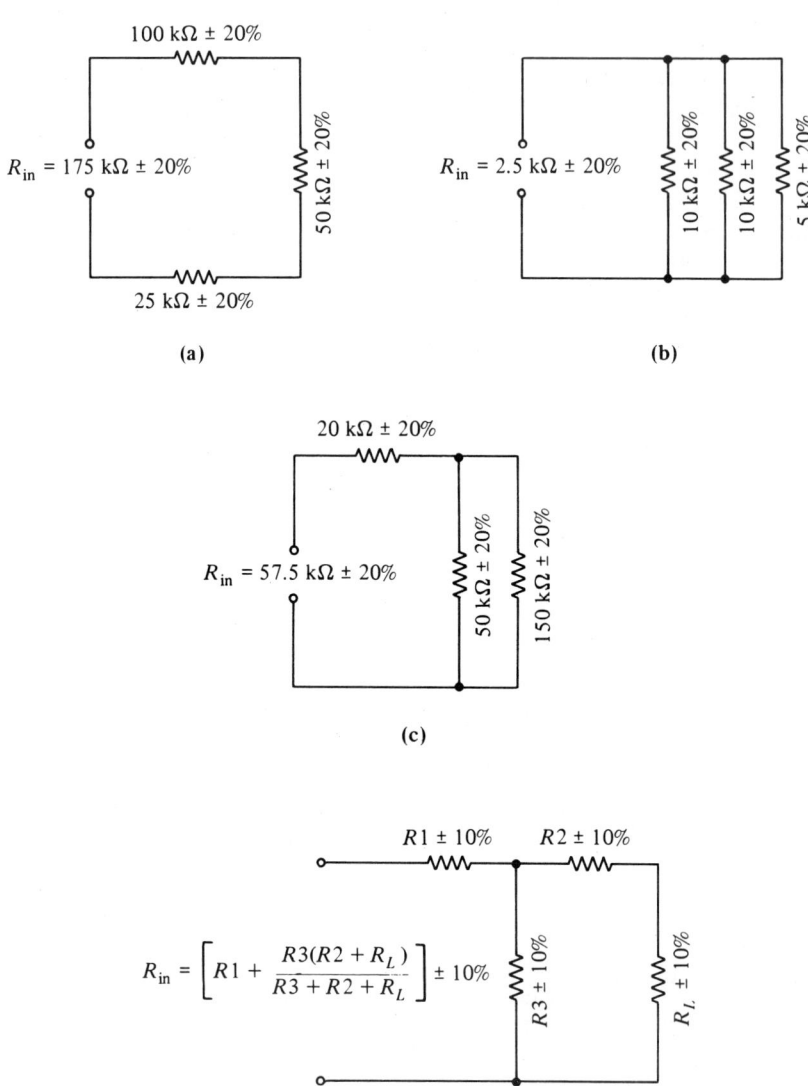

**Figure 1-5.** Resistors with equal tolerances combine to form a configuration with the same tolerance: (a) series configuration; (b) parallel configuration; (c) series–parallel configuration; (d) T-section configuration.

of 20 kΩ ± 15 percent. Next, if a 100-kΩ ± 10 percent resistor is connected in series with a 10-kΩ ± 20 percent resistor, their net resistance will have a value of 110 kΩ ± 11 percent. In this situation, the significant principle is that the tolerance on the net value is dominated by the tolerance

## 1-2 Tolerance Requirements and Calculations

on the large-value resistor. As a rough rule of thumb, note that the resistor that has 0.1 the value of the other resistor affects the net tolerance by only 10 percent. Thus the tolerance on the small-value resistor can be substantially relaxed without greatly affecting the tolerance on the net value.

It is possible for positive and negative tolerances to exactly cancel each other. For example, if a 1,000-$\Omega$ resistor happens to be at its upper tolerance limit of 1,100 $\Omega$, and it is connected in series with a 1,000-$\Omega$ resistor that happens to be at its lower tolerance limit of 900 $\Omega$, their net resistance will have the bogie value of 2,000 $\Omega$. The probability that positive and negative tolerances will cancel each other is theoretically the same as that both resistors will happen to be at their high tolerance limit, or that both will happen to be at their low tolerance limit. Whether the worst-case condition is represented by the high limit or by the low limit depends upon the intended circuit function.

### Basic RC Circuit Tolerances

RC circuits are widely used in audio configurations. Capacitive tolerances combine in the same manner as resistive tolerances. However, a different situation is encountered in the combination of resistive tolerances with capacitive tolerances, as exemplified in Fig. 1-6a. In this example, the bogie time constant of the RC circuit is 1 second (s). If the capacitor has a tolerance of $\pm 10$ percent, and the resistor has a tolerance of $\pm 10$ percent, the time constant of the circuit will have a tolerance of $+21$ percent and $-19$ percent. Because the resistance value is multiplied by the capacitance value, the resulting tolerance on the time constant is greater than the tolerance on the individual components. It is also for this reason that the time-constant value has a positive tolerance that is greater than its negative tolerance.

The effect of tolerances on the impedance of an RC circuit is depicted in Fig. 1-6b. If the resistance and the capacitive reactance have equal values, for example, a $\pm 10$ percent tolerance on these values causes the circuit impedance to have a $\pm 14$ percent tolerance. If a $\pm 10$ percent tolerance is assigned to the capacitance value (instead of the capacitive-reactance value), the resulting tolerance on the circuit impedance becomes slightly greater, owing to the inverse relationship between capacitance and capacitive–reactance values. A related basic consideration is the effect of tolerances on the phase angle of an RC circuit. As exemplified in Fig. 1-7, a tolerance of $\pm 10$ percent on the $R$ and $C$ values results in a tolerance of $\pm 5$ percent on the phase angle of the series RC circuit.

Consider next the effect of tolerances on the frequency response of an RC-coupled stage, such as shown in Fig. 1-8. Since the FET has a very high input impedance, the frequency response of the input circuit is de-

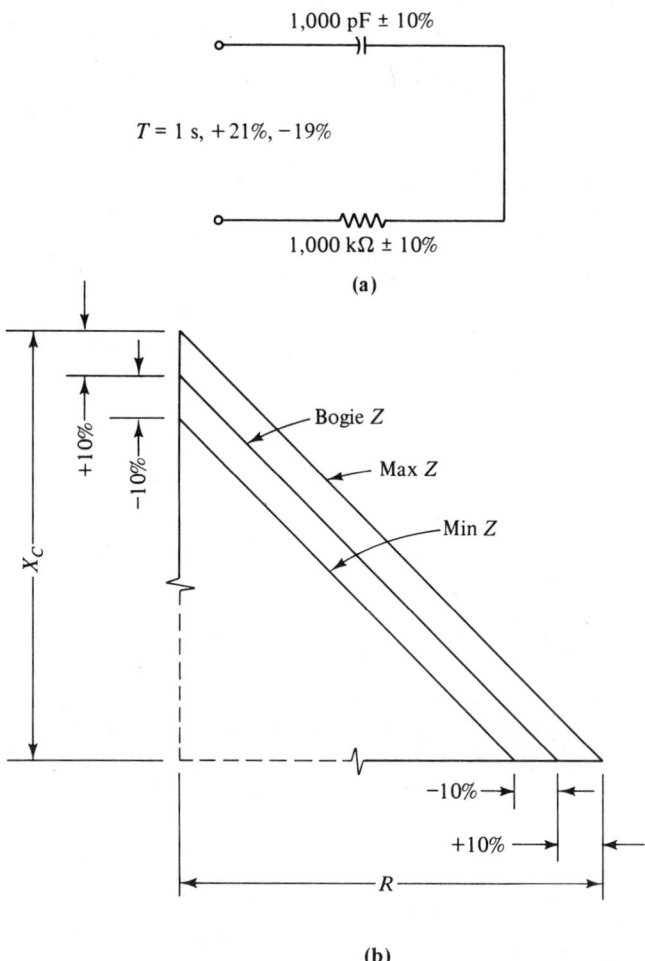

**Figure 1-6.** Effect of ±10 percent RC tolerances on time constant and impedance: (a) example of time-constant tolerance; (b) example of impedance tolerance.

termined by the values of $C$ and $R$. Bogie values in this example are 0.1 microfarad ($\mu$F) and 47 k$\Omega$. These values provide a $-$ 3-decibel (dB) low-frequency cutoff point at point 2 in Fig. 1-9. If the capacitor and the resistor have ± 20 percent tolerances, and both have their low-tolerance limiting values, the $-$3-dB cutoff point occurs at point 3. On the other hand, if the capacitor and resistor both have their high-tolerance limiting values, the $-$3-dB cutoff point occurs at point 1. From the viewpoint of the audio-amplifier designer, point 3 represents the worst-case situation. Note that the time constant of the emitter return circuit is greater than that

## 1-2 Tolerance Requirements and Calculations

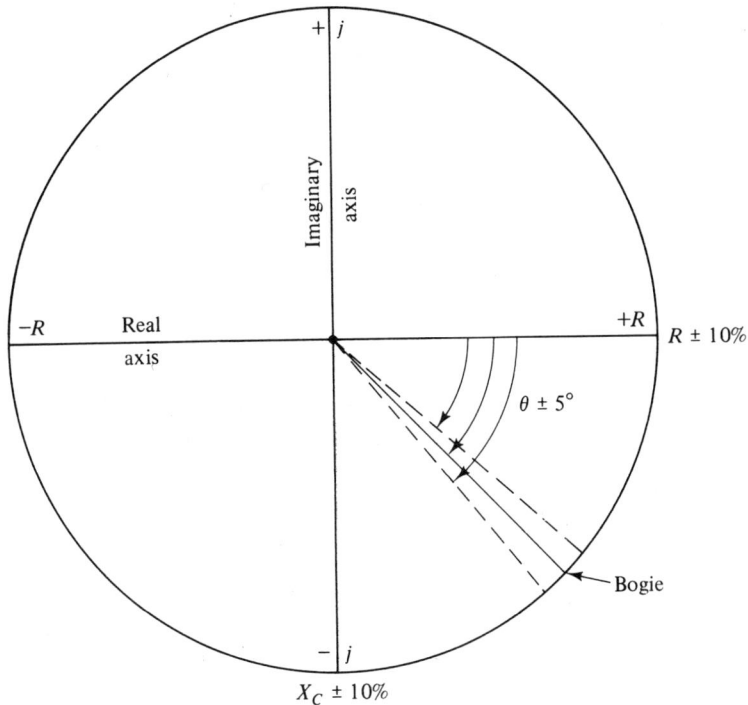

**Figure 1-7.** Tolerances on RC phase angle.

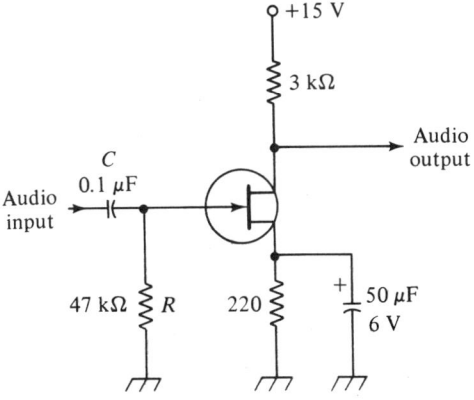

**Figure 1-8.** RC-coupled FET stage.

of the gate input circuit in this example. Accordingly, the low-frequency response of the stage is determined by the $R$ and $C$ values in the gate circuit, for all practical purposes.

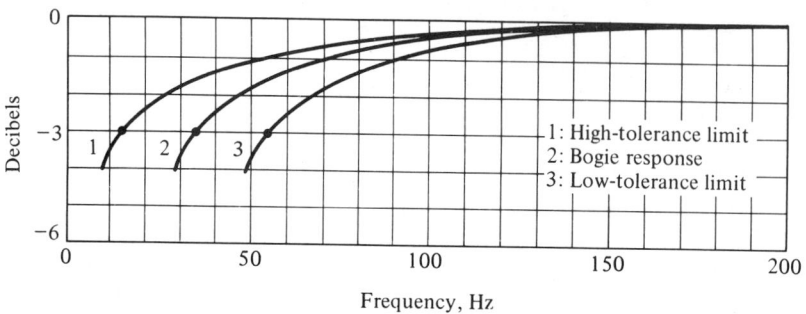

**Figure 1-9.** Tolerances on RC coupling-circuit frequency response.

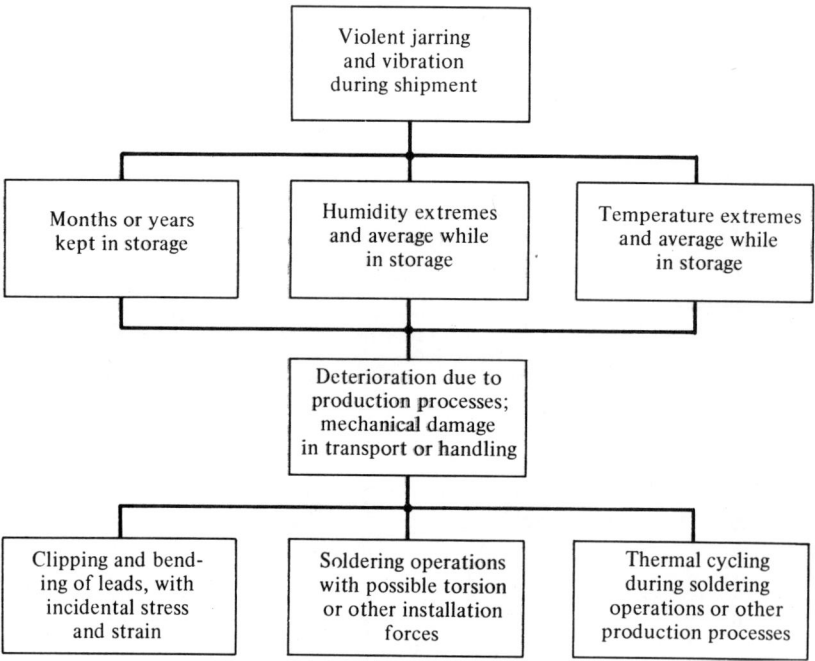

**Figure 1-10.** Basic causes for tolerance drift in resistors.

The audio circuit designer ordinarily assumes that all the resistors in a production lot are within their rated tolerance. This assumption is valid to a very high probability at the time of shipment. On the other hand, it is not necessarily true that all the resistors in the production lot will be found within rated tolerance after subjection to extended aging, after the leads have been clipped, bent, torsioned, soldered into a circuit board, and thermally cycled numerous times at various temperatures (see Fig. 1-10). Because production operations can affect the tolerance rating, the audio

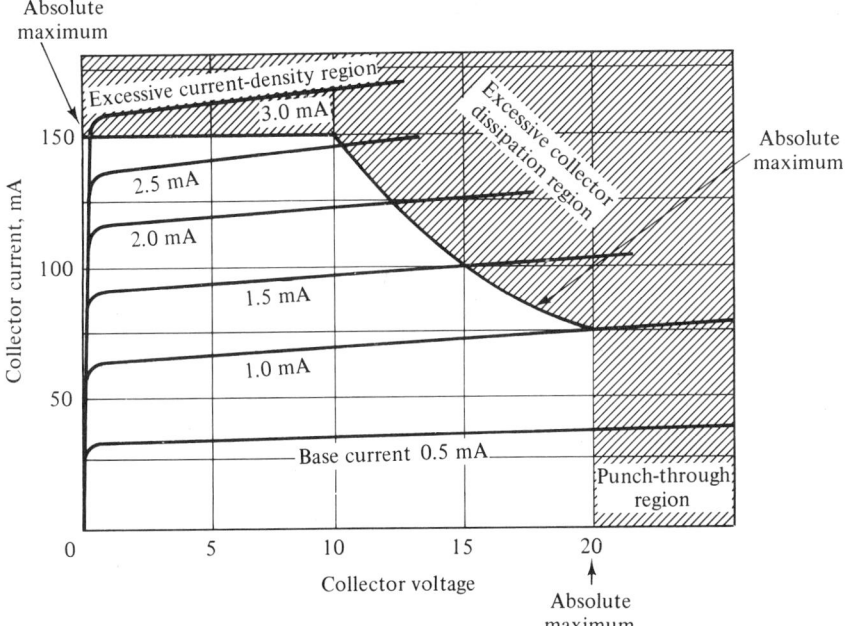

**Figure 1-11.** Forbidden regions of transistor operation.

circuit designer may be required to establish production test procedures to weed out resistors (and other components and devices) that have drifted out of tolerance.

## Worst-Case Considerations

*Worst-case circuit analysis* is defined as a design procedure that is utilized to determine the worst possible effect on the output parameters of the circuit owing to changes in the values of the circuit elements. Accordingly, the circuit elements are set to the values within their anticipated ranges that produce the maximum detrimental change in the output. Apart from rated tolerances, environmental factors such as temperature may need to be taken into account for worst-case values. As an illustration, the temperature coefficient of resistance could be a consideration in the analysis.

*Worst-case design* is defined as an extremely conservative design procedure wherein the circuit is planned to function normally, even though all the component and device values have simultaneously assumed the worst possible operating condition resulting from tolerance limits, environmental factors, and a stipulated degree of deterioration, such as would result from a period of aging. If the supply voltage has been assigned tolerance limits, this is also considered. Note that the power-supply voltage must never exceed an upper limit that is determined by the absolute-maximum rating of the transistor(s). Otherwise, device destruction can be anticipated. Forbidden regions of transistor operation are exemplified in Fig. 1-11.

## 1-3  Basic Device Tolerances

Device tolerances generally differ from component tolerances in that the former are seldom stated as a percentage "spread." Instead, a bipolar transistor, for example, is ordinarily rated for minimum current gain (beta), maximum collector cutoff current, maximum emitter cutoff current, minimum alpha cutoff frequency, and maximum output capacitance. Thus the audio circuit designer is justified in assuming that all transistors in a production lot will perform at least as well as rated, although an occasional transistor might have twice as much current gain as rated, for example. Low-level transistors are generally rated for maximum *noise figure,* such as 6 dB. The noise figure is defined as the ratio of the total noise power at the output of the transistor to that portion of the total output noise power attributable to the thermal agitation in the resistance of the signal source (typically 1,000 Ω).

Field-effect transistors (Fig. 1-12) are generally rated for minimum

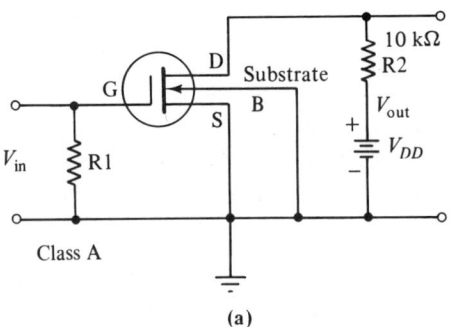

Voltage gain = 50 times
Transconductance: 5000 μhos
Power gain: 17 dB (50 times)
Input resistance: Very high
Output resistance: 20 kΩ
(For generator internal resistance of 500 Ω)

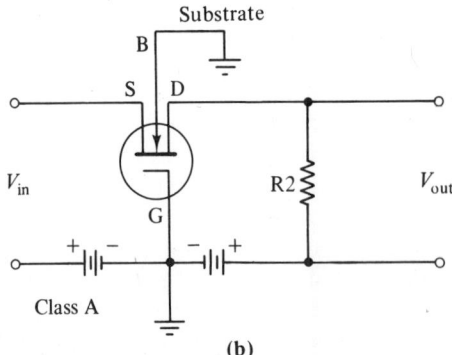

Voltage gain: 1.8
Input resistance: 240 Ω
Output resistance: High
(For generator internal resistance of 500 Ω)

**Figure 1-12.** Typical characteristics for unipolar transistor amplifier configurations: (a) common source; (b) common gate.

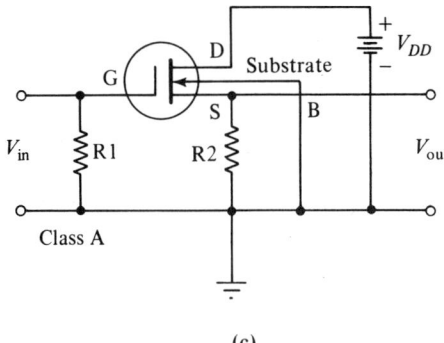

Voltage gain: 0.5
Input resistance: 2 MΩ
Output resistance: 240 Ω
(For generator internal resistance of 500 Ω)

Class A

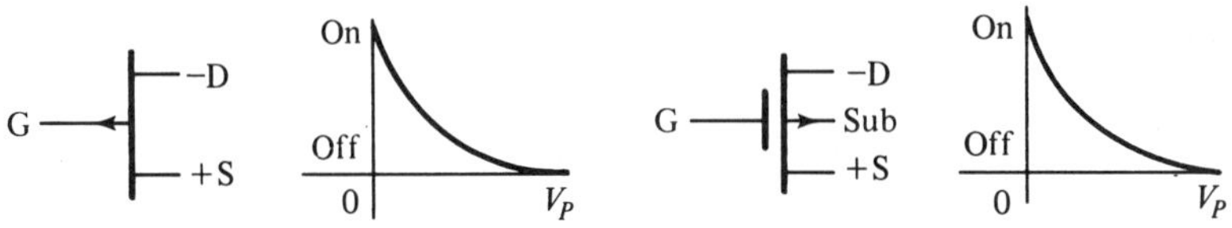

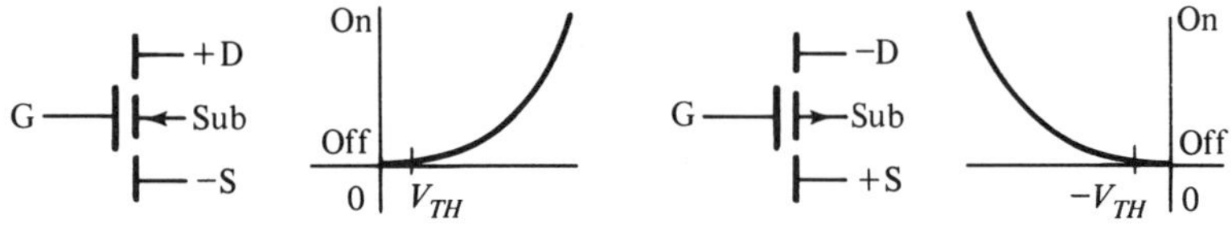

(c)

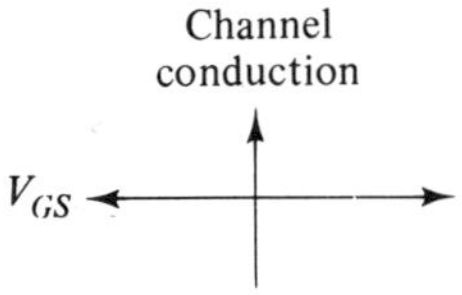

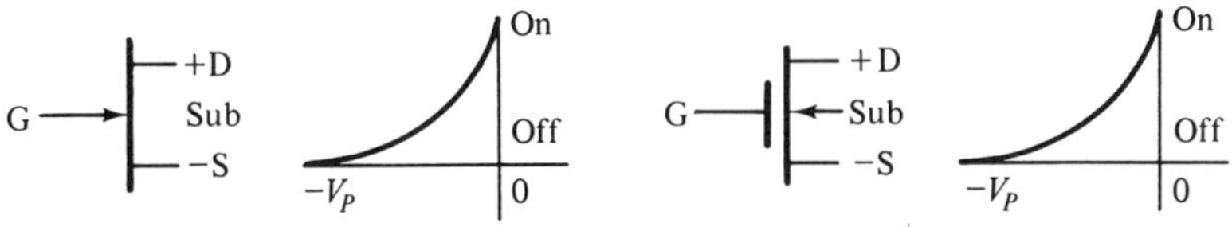

**Figure 1-12.** (*Continued*) (c) common drain; (d) N-channel JFET (depletion); (e) N-channel MOSFET (depletion); (f) P-channel JFET (depletion); (g) P-channel MOSFET (depletion); (h) N-channel MOSFET (enhancement); (i) P-channel MOSFET (enhancement).

power gain, such as 15 dB, and for typical power gain, such as 17 dB. Similarly, FET's are rated for a maximum noise figure, such as 6 dB, and for a typical noise figure, such as 4 dB. Because an FET is a voltage-operated device, a transconductance rating is usually assigned; for example, an FET may be rated for a minimum transconductance of 3,000 microsiemens ($\mu$S), and a typical transconductance of 6,000 $\mu$S. Note in passing that, although the operating characteristics of both bipolar and unipolar transistors have rather wide tolerances in production lots, the circuit designer can effectively tighten the device tolerances by use of negative feedback. As detailed subsequently, if a large amount of negative feedback is employed in an amplifier circuit, the input–output relations change but slightly as device characteristics are varied over a wide range (see Fig. 1-13).

Integrated circuits, as exemplified in Fig. 1-14, are generally rated for minimum and typical *open-loop gain*. Open-loop gain is defined as the ratio of the (loaded) output of the amplifier without any feedback to its net input at any frequency; voltage gain is usually stated. For example, an integrated circuit utilized in the configuration of Fig. 1-14 is rated for a minimum open-loop gain of 53 dB and a typical open-loop gain of 58 dB. An integrated circuit is also rated for open-loop −3-dB bandwidth; in this example, the IC is rated for a minimum value of 250 kilohertz (kHz) and a typical value of 300 kHz. This IC is rated for a maximum noise figure of 2 dB at 1 kHz. It is also rated for a minimum output voltage swing of 2 V at 1 kHz, with a total harmonic distortion of 5 percent or less. Its typical output-voltage swing at 5 percent distortion is rated at 2.4 V. The audio circuit designer employs negative feedback to reduce harmonic distortion as much as may be desired.

## 1-4 Principles of Power Dissipation

A resistor is always rated for maximum power dissipation at room temperature. Most resistors are also rated for maximum voltage drop. As an illustration, if a 1-megohm (M$\Omega$) resistor has 1-watt (W) construction and a 600-V maximum rating, it cannot dissipate more than 0.36 W without exceeding its voltage rating. If its voltage rating is exceeded, the resistor is likely to be short-lived and to arc fail. Whenever a resistor is operated at an elevated ambient temperature, it should be derated accordingly; if it is operated at a reduced ambient temperature, it may be operated at a power level that exceeds its nominal dissipation rating. A derating diagram for a metal-film resistor is exemplified in Fig. 1-15. The circuit designer selects a value of ambient temperature along the horizontal axis

## 1-4 Principles of Power Dissipation

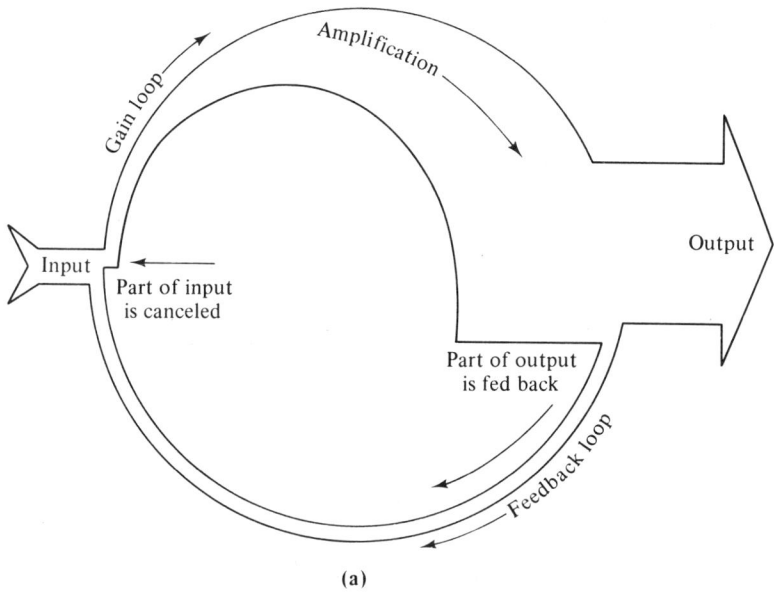

(a)

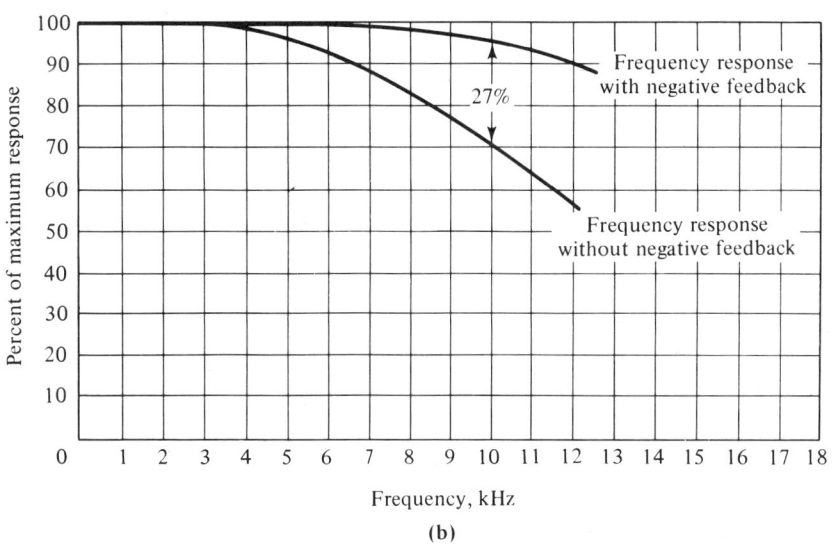

(b)

**Figure 1-13.** Negative-feedback action: (a) plan of a negative-feedback system; (b) improvement of frequency response by 20 dB of negative feedback.

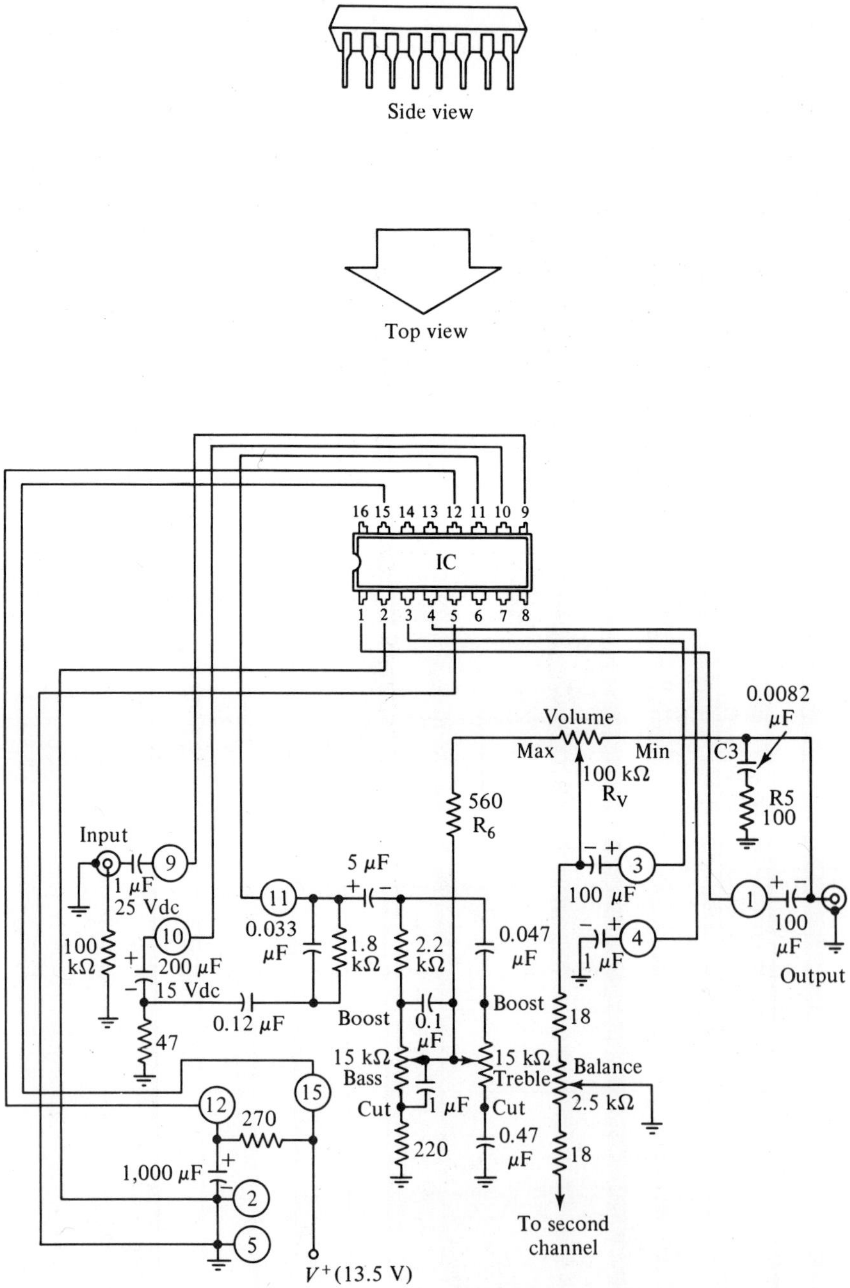

**18**  Overview of Basic Design Principles

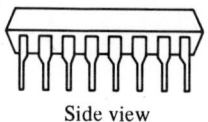
Side view

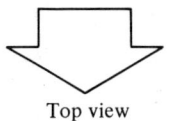
Top view

**Figure 1-14.** Integrated-circuit preamplifier arrangement.

## 1-4 Principles of Power Dissipation

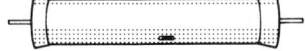

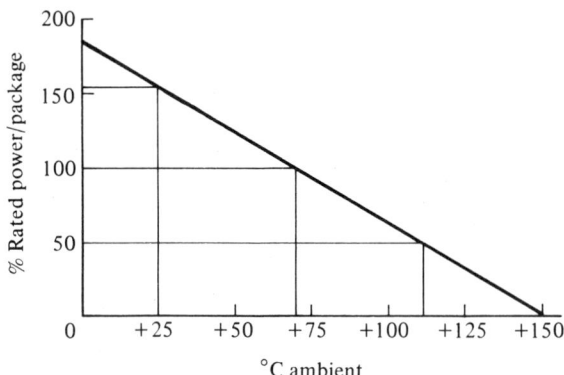

**Figure 1-15.** Typical derating diagram for a metal-film resistor.

and erects a perpendicular line to the "hypotenuse" in the diagram. This point of intersection is projected to the ordinate, which is calibrated in percentage units. Thus, if the projection occurs at the 50 percent point on the ordinate, the circuit designer will derate the resistor to one half of its rated power dissipation. On the other hand, if the projection occurs at the 155 percent point on the ordinate, he will assign a value of 155 percent to the rated power dissipation of the resistor.

If resistors are connected in series, the permissible voltage drop of the combination is increased, and the permissible power dissipation of the combination is also increased. For example, if two 100,000-Ω 1-W resistors are connected in series, the maximum power dissipation of the combination becomes 2 W. Similarly, if each resistor has a 300-V maximum rating, the series combination will have a 600-V maximum capability. On the other hand, if a 150,000-Ω 1-W resistor is connected in series with a 50,000-Ω 1-W resistor, the power-dissipation capability of the combination is *not* 2 W. Since both resistors carry the same current, the power dissipated by the 150,000-Ω resistor is three times the power that is dissipated by the 50,000-Ω resistor. Or the maximum power capability of this series combination is 1.33 W. The series-connected resistors operate as a voltage divider, and the voltage drops across the individual resistors are proportional to their resistance values. In other words, the 150,000-Ω resistor drops 75 percent of the applied voltage, and the 50,000-Ω resistor drops 25 percent of the applied voltage.

When resistors are connected in parallel, the permissible voltage drop for the combination is the same as that of the resistor with the lowest voltage rating. On the other hand, the permissible power dissipation of the combination is greater than that of any individual resistor. As an illustration, if two 50-Ω 5-W resistors are connected in parallel, their 25-Ω combination has a power-dissipation capability of 10 W. However, if a 30-Ω 5-W resistor is connected in parallel with a 150-Ω 5-W resistor, this parallel combination does *not* has a power-dissipation capability of 10 W. Inasmuch as both resistors have the same voltage drop, the 30-Ω resistor dissipates five times as much power as the 150-Ω resistor; thus the maximum power capability of this parallel combination is only 6 W.

A potentiometer is rated for maximum power dissipation, such as 1 W. This rating applies to the total resistance element. For example, if a 1-W potentiometer has a total resistance of 2,500 Ω, the circuit designer may apply 50 V across the 2,500-Ω element. On the other hand, if the potentiometer is set to its 500-Ω position, it can dissipate only 0.2 W, and a maximum of 10 V can be safely applied across the 500-Ω section of the potentiometer. Consider next the circuit shown in Fig. 1-16. In this example, the potentiometer operates into a comparatively low resistance load. If the potentiometer is rated for a maximum power dissipation of 0.25 W, this rating will be exceeded at any position other than the minimum setting. For example, if the potentiometer is set to its midpoint, the upper half of the resistance element must dissipate practically 0.5 W. Therefore, the potentiometer must be chosen with a suitable power-dissipation rating.

Audio potentiometers are available with various *tapers,* as shown in Fig. 1-17. Circuit designers employ nonlinear tapers in nonlinear configurations. That is, a potentiometer taper is chosen that is approximately the inverse of the configuration taper. As a result, the net control action is

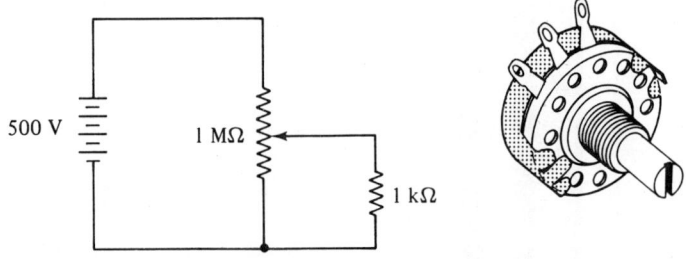

**Figure 1-16.** Potentiometer circuit with a comparatively low resistance load.

## 1-4 Principles of Power Dissipation

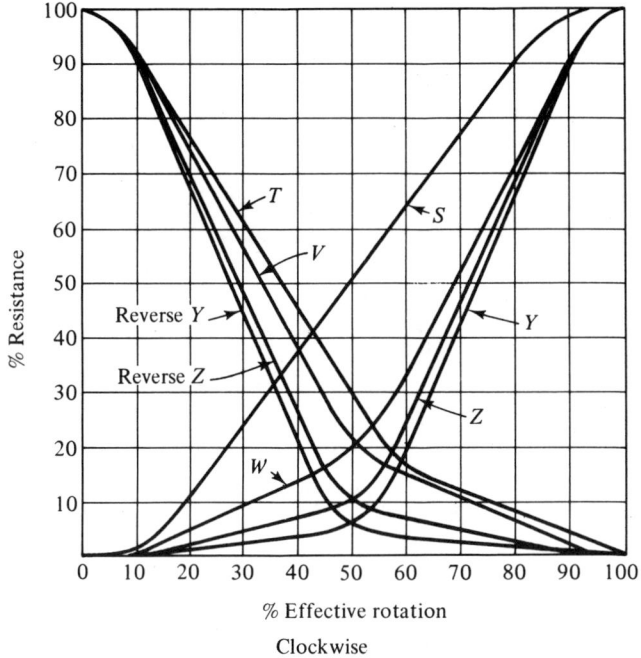

Figure 1-17. Standard potentiometer tapers for audio circuitry.

Taper S — Straight or uniform resistance change with rotation.
Taper T — Right-hand 30% resistance at 50% of counterclockwise rotation.
Taper V — Right-hand 20% resistance at 50% of counterclockwise rotation.
Taper W — Left-hand 20% resistance at 50% of clockwise rotation.
Taper Z — Left-hand (log. audio) 10% resistance at 50% clockwise rotation.
Rev. Z — Right-hand 10% resistance at 40% counterclockwise rotation.
Taper Y — Left-hand 5% resistance at 50% of clockwise rotation.

approximately proportional to the number of degrees of potentiometer rotation. This relation makes the setting of the control less critical, and provides a more logical equipment response from the viewpoint of the user. Note that circuits comprising resistance, capacitance, and inductance are basically linear. On the other hand, circuits that include solid-state devices may have nonlinear input–output relationships. For example, a transfer characteristic (base voltage versus collector current) for a small-signal transistor is shown in Fig. 1-18. Another factor that bears on taper

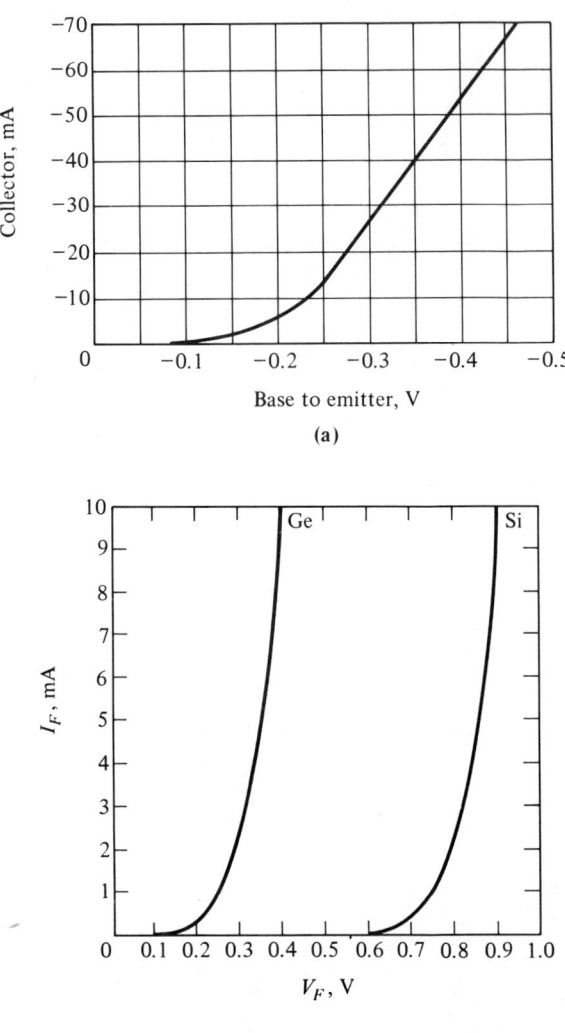

**Figure 1-18.** Transfer characteristics for a typical small-signal transistor: (a) bias voltage versus collector current for a germanium transistor; (b) comparison of bias voltages for silicon and germanium transistors.

considerations is the loudness characteristic of the hearing process. This characteristic is measured in phon units, as detailed subsequently in Section 1-10.

Transistors are rated for maximum power dissipation at room temperature. For example, a typical audio power transistor is rated for a maximum dissipation of 10 W. The dissipation is equal to the product of collector voltage and collector current in class A operation. As detailed subsequently, the dissipation can also be calculated in a somewhat similar manner for class AB and class B operation. The maximum case temperature for the transistor at maximum rated power dissipation, in this example, is 80°C. In turn, if this transistor is to be operated with a case temperature of 150°C, its maximum rated power value must be derated. With reference to Fig. 1-19, a perpendicular line is drawn from the 150°C point on the abscissa to the 200°C (maximum operating temperature) line. In turn, the projection of this point to the ordinate intersects at the 30 percent point on the ordinate. Accordingly, the maximum permissible power dissipation for the exemplified transistor at a case temperature of 150°C (centigrade) is 3 W.

## 1-5 Law of Probability

Various audio circuit design considerations are based on laws of probability. To cite a very simple example, suppose that there are eight high-value resistors and two low-value resistors in a lot. If a resistor is selected

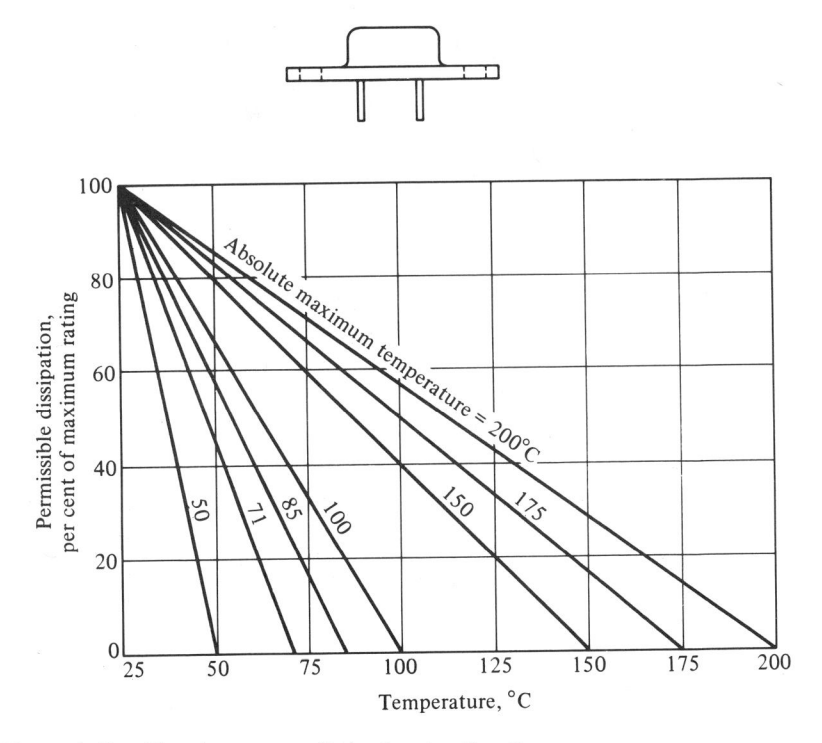

**Figure 1-19.** Transistor power-dissipation derating diagram.

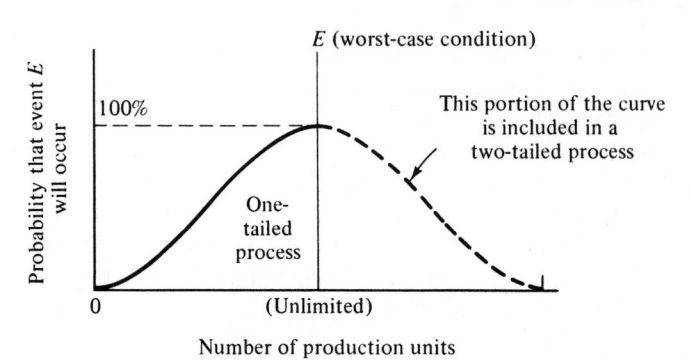

**Figure 1-20.** Qualitative example of the probability of a worst-case occurrence.

at random, the assembler is four times more likely to pick up a high-value resistor than to pick up a low-value resistor. Otherwise stated, the probability that he will pick up a high-value resistor is $8/10$, or $4/5$. Or the odds are 4 to 1 that the resistor will have a high value. The law of normal distribution defines the range of probabilities for a random occurrence. For example, an extreme worst-case occurrence in a production operation is a random occurrence that has a certain probability. That is, it is not highly probable that all component and device tolerances will happen to have their worst-case values in a random unit of equipment. Nevertheless, this possibility exists, and it is sometimes feasible to calculate the probability of a worst-case occurrence in a production process.

The audio circuit designer may attempt to specify component and device tolerances which ensure that an occasional worst-case occurrence will still fall within acceptable performance limits. More commonly, the designer chooses to relax component and device tolerances from an economical viewpoint, and to take a calculated risk that the number of production rejects will not increase overall manufacturing costs objectionably. A standard probability curve is shown in Fig. 1-20. This example indicates in a qualitative manner how the probability of a worst-case occurrence varies versus the number of units produced. In other words, if an unlimited (theoretically infinite) number of units are produced, there is a 100 percent probability of a worst-case occurrence. Conversely, if very few units are produced, the probability of a worst-case occurrence is very small. This is an elementary statement of a one-tailed process wherein only one half of the total probability curve is applicable. If we consider the distribution of high- and low-value resistors in a random production lot, for example, a two-tailed analysis is employed wherein the total probability curve applies.

Worst-case occurrences in a production operation can be analyzed in terms of standard probability calculations only when the component

## 1-6 Audio Design Pitfalls

and device tolerances are truly random. It often happens, in practical situations, that tolerances in a parts shipment tend to cluster toward the low end or toward the high end of the specified limits. Although clustering can be taken into account in the probability calculations, at least in a general way, design engineers seldom take this complication into account, and assume that component and device tolerances are truly random. Technically, the *probability distribution* is a mathematical model that shows a representation of the probabilities for all possible values of a given random variable. The *probability of success* is defined as the likelihood that a production unit will function satisfactorily for a stated period of time when subjected to a specified environment.

## 1-6 Audio Design Pitfalls

Experienced circuit designers know that unexpected pitfalls may be encountered in various phases of a design project. For example, unless amplifiers are adequately decoupled, they may become unstable when connected to a common power supply. Instability may impair the system frequency response and cause distortion; in severe cases, *motorboating* (self-oscillation) can occur. A typical power-supply configuration for an audio system is shown in Fig. 1-21. Although the internal impedance of the power supply is small, it is not zero. As the filter capacitors C1, C2, and C3 gradually deteriorate, the internal impedance of the power supply increases. Note also that there is a small value of common impedance between outputs 1 and 2. Because of these residual power-supply impedances, an amplifier that operates satisfactorily from a bench power

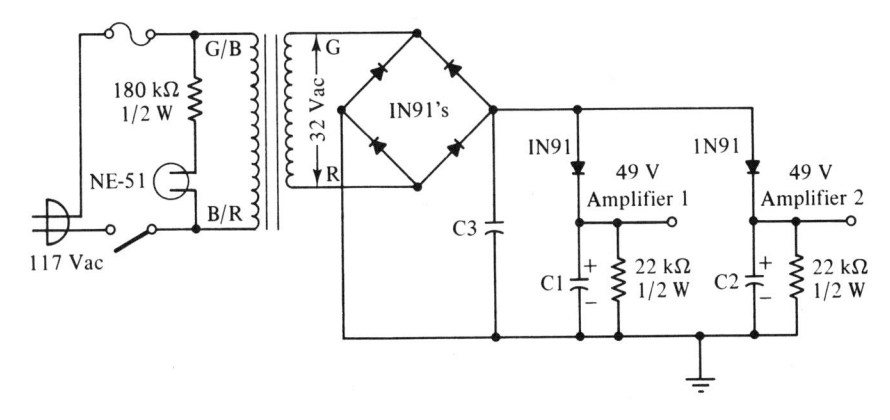

C1, C2, and C3: 1,500 μF, 50 V

Figure 1-21. Typical power supply for an audio system.

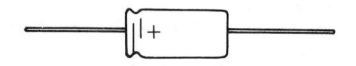

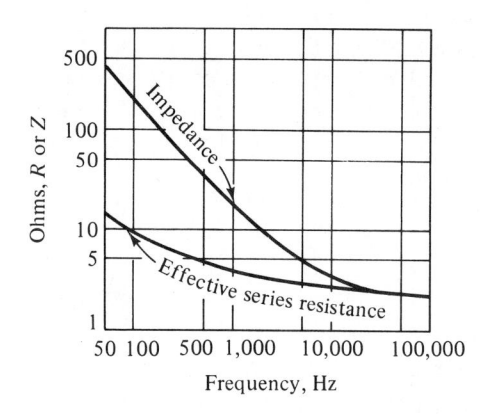

**Figure 1-22.** Frequency characteristics of a typical 8-$\mu$F electrolytic capacitor.

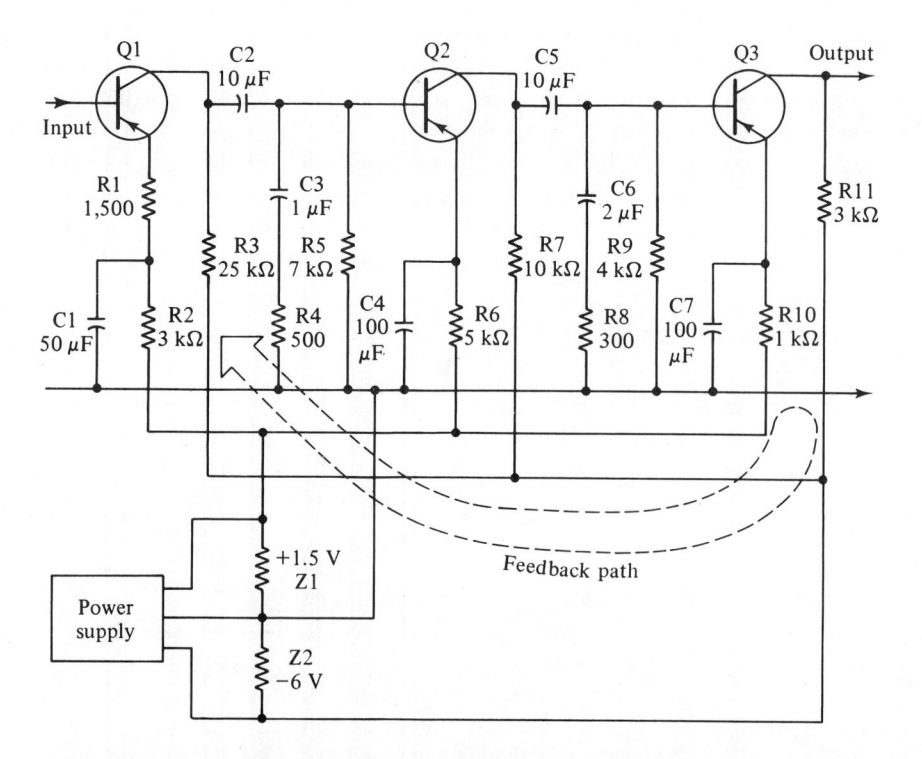

**Figure 1-23.** Example of feedback path in a common power-supply system.

supply may or may not remain stable when it is energized from a common power supply that also energizes other amplifiers in a system. The impedance of a typical 8-$\mu$F electrolytic capacitor is exemplified in Fig. 1-22. Therefore, the audio-circuit designer should be aware of this pitfall, and he should check any new amplifier arrangement under system conditions before it is released to production.

An example of a feedback path in a common power-supply system is shown in Fig. 1-23. A phase reversal occurs from base to collector of Q2, and another phase reversal occurs from base to collector of Q3. In turn, the common power-supply impedance Z2 feeds back in-phase energy from the collector of Q3 to the base of Q1. This is a regenerative condition that causes a peaked frequency response. This distortion occurs at the frequency for which the system within the feedback loop has maximum gain. If the value of Z2 exceeds a certain critical value, Q2 and Q3 will motorboat, and a putt-putt sound will be radiated from the speaker. Next, observe that emitter resistors R6 and R10 are effectively decoupled from the power-supply impedance by capacitors C4 and C7. Therefore, impedance Z1 can have a high value without development of a significant feedback path. To stabilize the collector-return circuits, the circuit designer includes an RC decoupling "network" in each collector-return lead. This "network" comprises a series resistor and a shunt capacitor.

If a prototype model operates within performance specifications, the designer may tend to relax vigilance and to assume that the circuit has been adequately verified. However, the careful designer will analyze the model for possible pitfalls that may have been overlooked. For example, all electronic equipment generates more or less heat. Heat sources are localized within a unit of equipment. The rise of internal operating temperature above ambient is difficult to calculate and must usually be determined experimentally. A critical designer checks equipment operation over an extended period in the highest specified ambient temperature. He measures the temperature at each component and device that is rated for a maximum operating temperature. If any measurement reveals that there is little or no margin between the rated maximum and the operating temperature (under worst-case conditions), the designer will modify the parts list and use a suitably derated component or device.

## 1-7 Underwriters' Laboratories (UL) Approval

Specific guidelines have been established by the Underwriters' Laboratories for minimizing shock and fire hazards in the design of electrical and electronic equipment. The Underwriters' Laboratories are a branch of the National Board of Fire Underwriters. A consumer electronics product that has been examined by the UL, and which has been accepted, is permitted to carry a tag or label of approval, as shown in Fig. 1-24. Circuit designers can obtain a pamphlet from any UL office to determine the prevailing electrical and mechanical requirements for approval by UL inspectors. Specific approval of each design and model is necessary. Many municipalities require UL approval for all products sold in their jurisdictional areas.

**Figure 1-24.** Typical Underwriters' Laboratories labels.

Some basic UL requirements are as follows. A power transformer used in a consumer-electronics product must be completely enclosed in a metal or other noncombustible housing to prevent the escape of flames or molten metal in case the transformer were destroyed by a catastrophic overload. A power transformer with a UL-approved housing is depicted in Fig. 1-25. A circuit breaker or fuse must be included in the input line of the power supply to minimize the danger of fire in the event of a serious overcurrent demand anywhere in the equipment circuitry. Temperature limits for various classes of materials must be observed. Each conductor is required to have adequate cross section and insulation for its intended current flow, operating temperature, and application. Power cords must meet specific requirements for stranded conductors at the rated current demand, and must be adequately insulated. A power cord must be provided with strain relief inside the equipment; the cord must enter through an insulating bushing that will prevent cutting or fraying

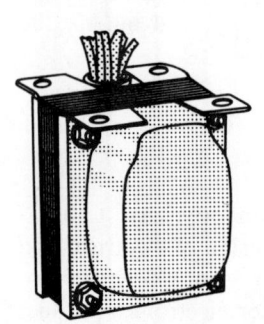

**Figure 1-25.** Power transformer with a UL-approved housing.

of the insulation on the cord. Also, the power cord must be secured at its entry point so that it cannot be pushed back into the interior of the cabinet.

## 1-8 Black-Box Concept

The audio-circuit designer often starts with a black-box concept, defined by performance specifications. This is a useful mathematical approach to an electronic configuration that concerns itself only with input and output parameters and relations, and ignores the interior elements, discrete or integrated. A black-box representation for a high-fidelity preamplifier is pictured in Fig. 1-26. The input and output data are basic design requirements. Next the audio circuit designer proceeds to break down the black box and to implement the basic performance specifications. For example, he will determine how many stages are required and decide on the types of devices that will be utilized. Then circuit details are worked out, and the design is breadboarded. If the breadboard model meets the black-box data, it is next constructed in prototype model form. Further tests are then made to ensure that the design is reproducible in large-scale production.

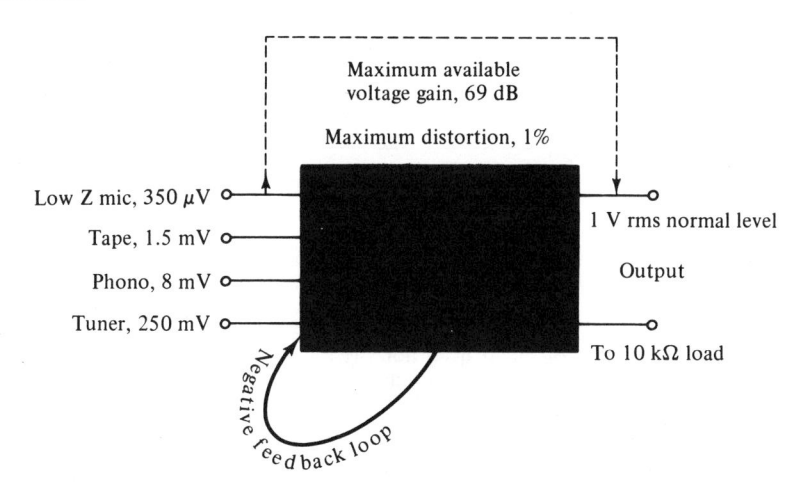

**Figure 1-26.** Example of an audio black-box concept.

## 1-9 Basic Types of Distortion

Five basic types of distortion include frequency distortion, phase distortion, amplitude distortion, transient distortion, and parasitic distortion, as exemplified in Fig. 1-27. Although there is a relation between frequency

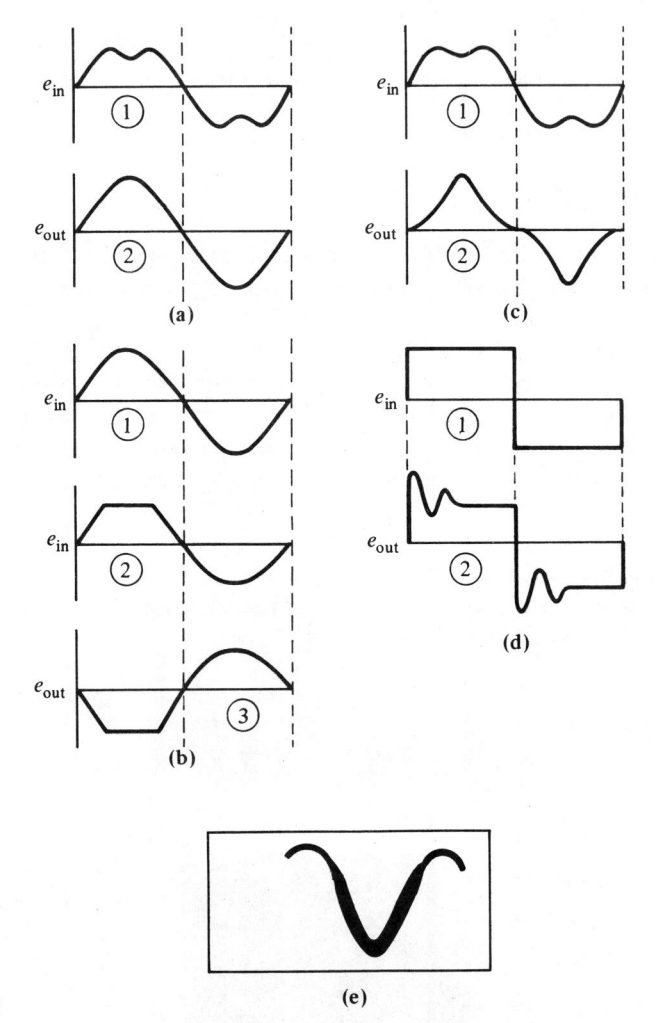

**Figure 1-27.** Basic types of audio distortion: (a) frequency distortion; (b) amplitude distortion; (c) phase distortion; (d) transient distortion; (e) parasitic distortion.

distortion and transient distortion, this relation is complex, and it is advisable for the audio circuit designer to regard them as separate categories. There are various forms of square-wave distortion, as shown in Fig. 1-28. High-fidelity components are usually tested with a square-wave repetition rate of 2 kHz, although other repetition rates may also be utilized. Crossover distortion is depicted in Fig. 1-29. It is essentially a form of amplitude distortion and produces harmonics in the output wave

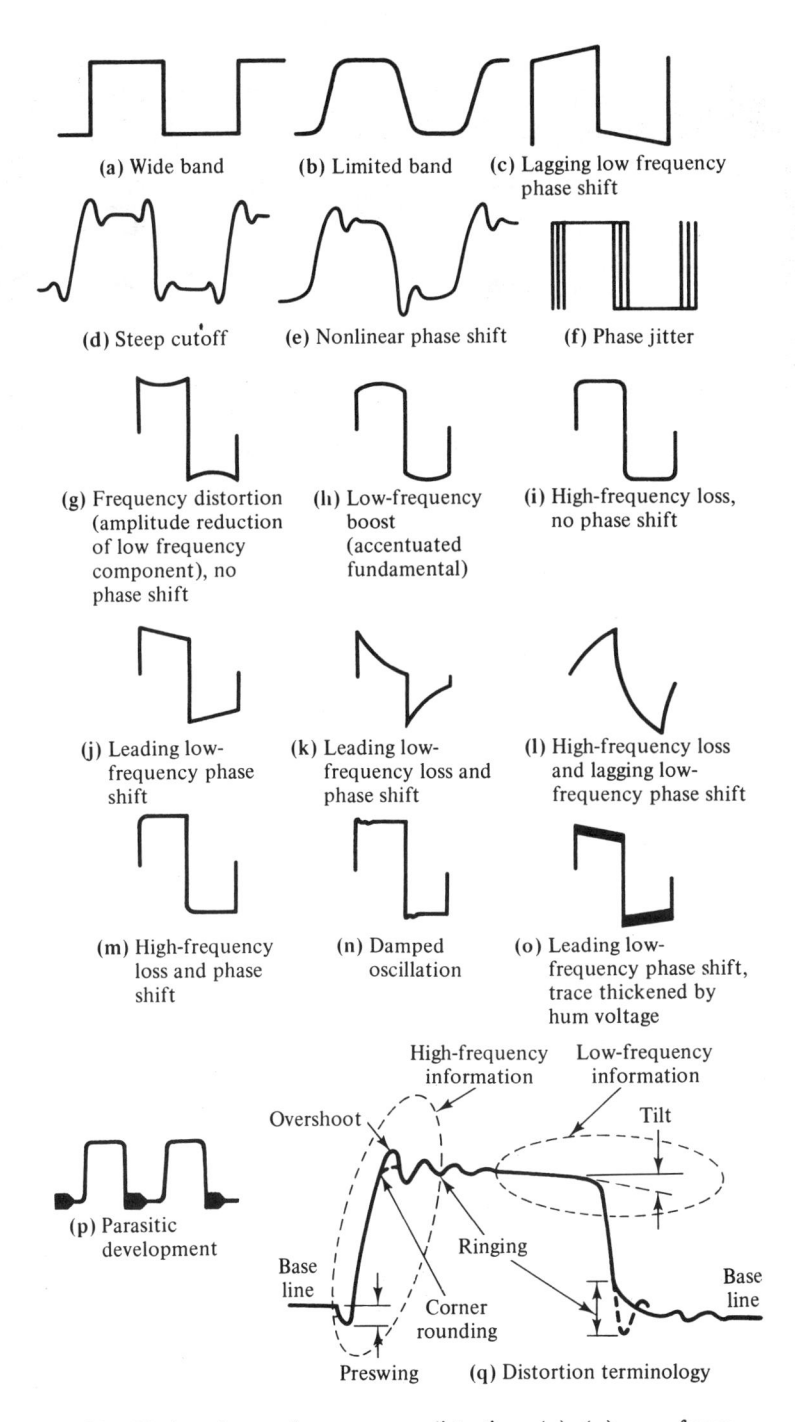

**Figure 1-28.** Various forms of square-wave distortion: (a)–(q) wave forms.

(a) Wide band
(b) Limited band
(c) Lagging low frequency phase shift
(d) Steep cutoff
(e) Nonlinear phase shift
(f) Phase jitter
(g) Frequency distortion (amplitude reduction of low frequency component), no phase shift
(h) Low-frequency boost (accentuated fundamental)
(i) High-frequency loss, no phase shift
(j) Leading low-frequency phase shift
(k) Leading low-frequency loss and phase shift
(l) High-frequency loss and lagging low-frequency phase shift
(m) High-frequency loss and phase shift
(n) Damped oscillation
(o) Leading low-frequency phase shift, trace thickened by hum voltage
(p) Parasitic development
(q) Distortion terminology — High-frequency information, Low-frequency information, Overshoot, Tilt, Ringing, Base line, Corner rounding, Preswing, Base line

31

**Figure 1-28.** (*Continued*) (r) a useful generator that provides square-wave output. (*Courtesy, Heath Co.*)

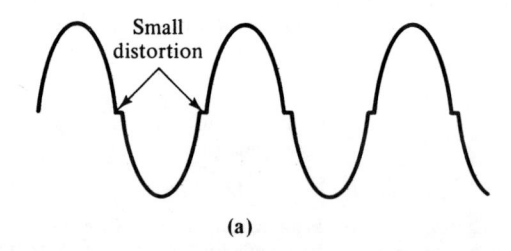

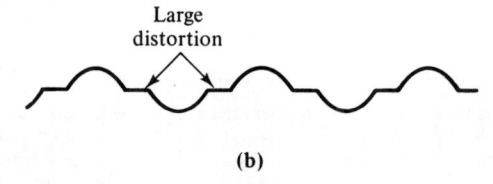

**Figure 1-29.** Examples of crossover distortion: (a) output at higher amplitude; (b) output at lower amplitude.

form. However, crossover distortion differs from clipping distortion, for example, in that the percentage of crossover distortion measured with a harmonic-distortion meter increases as the amplifier power output decreases. On the other hand, the percentage of clipping distortion increases as the amplifier power output increases.

Crossover distortion is typically produced by class B push–pull amplifiers when the transistors are operated with zero bias. Hence designers generally operate push–pull output stages in class AB, with a small amount of forward bias on the transistors. If excessive forward bias is employed in class AB operation, stretching distortion occurs. When one

## 1-9 Basic Types of Distortion

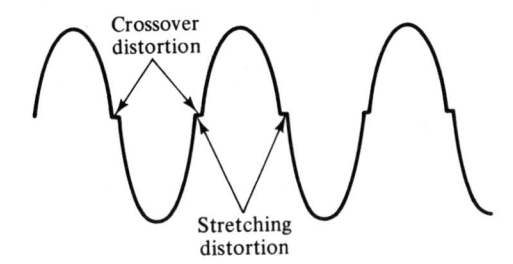

**Figure 1-30.** Alternate crossover and stretching distortion.

of the output transistors has insufficient forward bias and the other transistor has excessive forward bias, alternate crossover and stretching distortion will result, as shown in Fig. 1–30. When an excessive amount of incorrectly designed negative feedback is used in an audio amplifier, transient intermodulation distortion results, as depicted in Fig. 1-31.

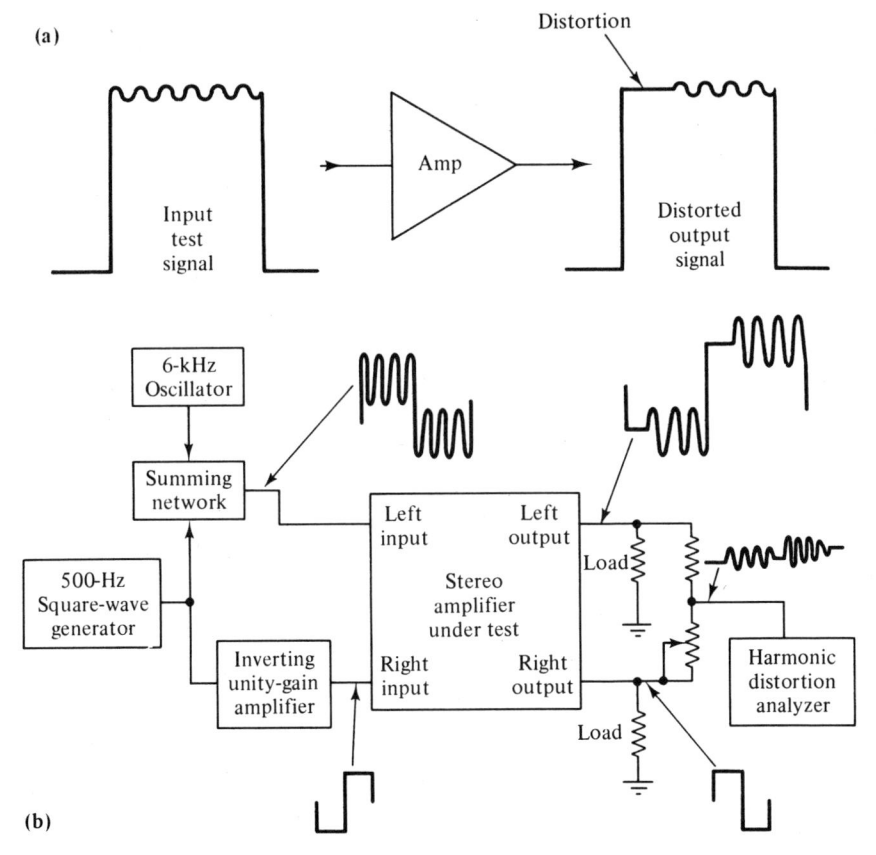

**Figure 1-31.** Example of transient intermodulation distortion (TIM): (a) input and output wave forms; (b) test arrangement for check of transient intermodulation distortion.

## 1-10 Phon and Mel Units

The phon is a loudness unit. Thus it is distinguished from the decibel referred to 1 milliwatt (dBm), which is a power unit. The loudness unit is of basic importance in audio systems, because the ear is not equally responsive to a given power level at various frequencies. Thus the phon is a function of frequency and represents different power levels at different frequencies; a phon unit has the same loudness at any chosen frequency. The phon relation to frequency and to decibels is depicted in Fig. 1-32. By definition, a frequency of 1 kHz is taken as a common reference point, so that at this frequency the phon level is equal to the decibel level (provided that the same reference level is used for both measurements). Zero reference level for the loudness unit is standardized at 0.0002 dyne/square centimeter ($cm^2$). With reference to Fig. 1-32, a loudness control in a high-fidelity amplifier is frequency compensated to conform to the curves of constant-phon values at corresponding power levels.

The phon unit should not be confused with the *mel* unit. In other words, the phon is a unit of loudness, whereas the mel is a unit of pitch. A pure 1-kHz sine wave, 40 dB above a listener's threshold, produces a pitch of 1,000 mels as perceived by the listener. The pitch of any sound

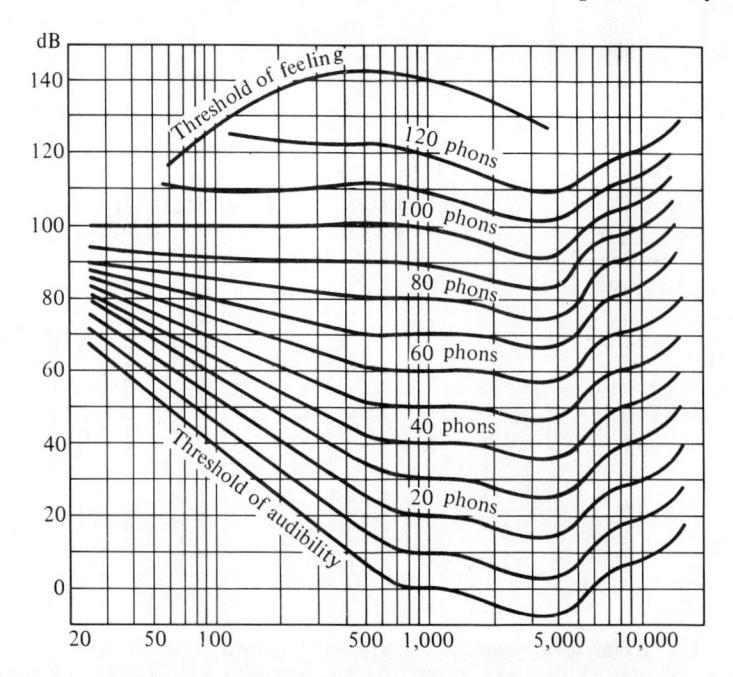

**Figure 1-32.** Relation of loudness units (phons) to frequency and to decibels.

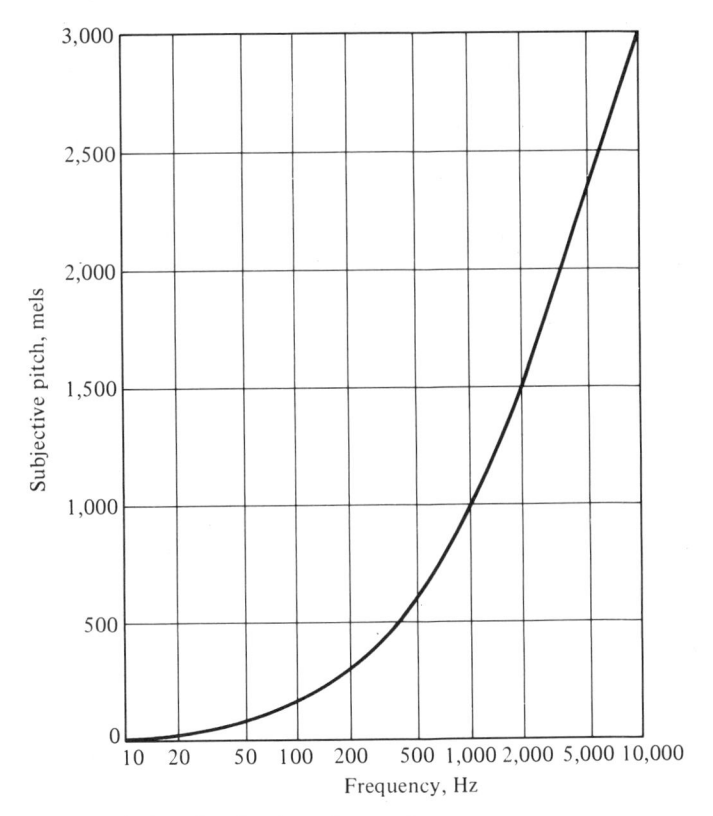

**Figure 1-33.** Relation of mel units to hertz units.

that is judged by the listener to be $n$ times that of a 1-mel pitch has a value of $n$ *mels*. Like the phon, the mel is a subjective unit, whereas the decibel is a physical unit. The relation of the mel scale to frequency is shown in Fig. 1-33. Observe that if 1,000 mels is taken as a reference frequency, it corresponds to a frequency of 1,000 hertz (Hz). Next, a pitch of 2,000 mels corresponds unexpectedly to a frequency of 4,000 Hz. Again, 500 mels corresponds unexpectedly to a frequency of 400 Hz. These examples point up the fact that perception of sound is a nonlinear type of response.

## 1-11 Low-Frequency Boost

When the audio circuit designer desires to obtain extended low-frequency response from an RC-coupled amplifier, a reasonable extension can be obtained by including a low-frequency boost circuit in the first stage of the amplifier, as shown in Fig. 1-34. When suitable values of boost components, $R_B$ and $C_B$, are utilized, the low-frequency response can be appreciably improved. In this example, the first screen photo shows the 60-Hz square-wave response of the uncompensated amplifier. Next, 60-Hz square-wave responses for various values of $R_B$ and $C_B$ are illustrated. It is seen that a choice of 15 kΩ and 2 μF provides optimum 60-Hz square-wave response in this example. Note that the required values of $R_B$ and $C_B$ may be rather critical in a multistage arrangement; also, these values depend considerably upon the tolerance deviations of the R and C coupling networks in the various stages.

(a)

Left to right, top: Uncompensated, 5 kΩ and 2 μF, 5 kΩ and 1 μF, 10 kΩ and 1 μF; left to right, bottom: 10 kΩ and 2 μF, 15 kΩ and 1 μF, 15 kΩ and 2 μF, 20 kΩ and 1 μF

(b)

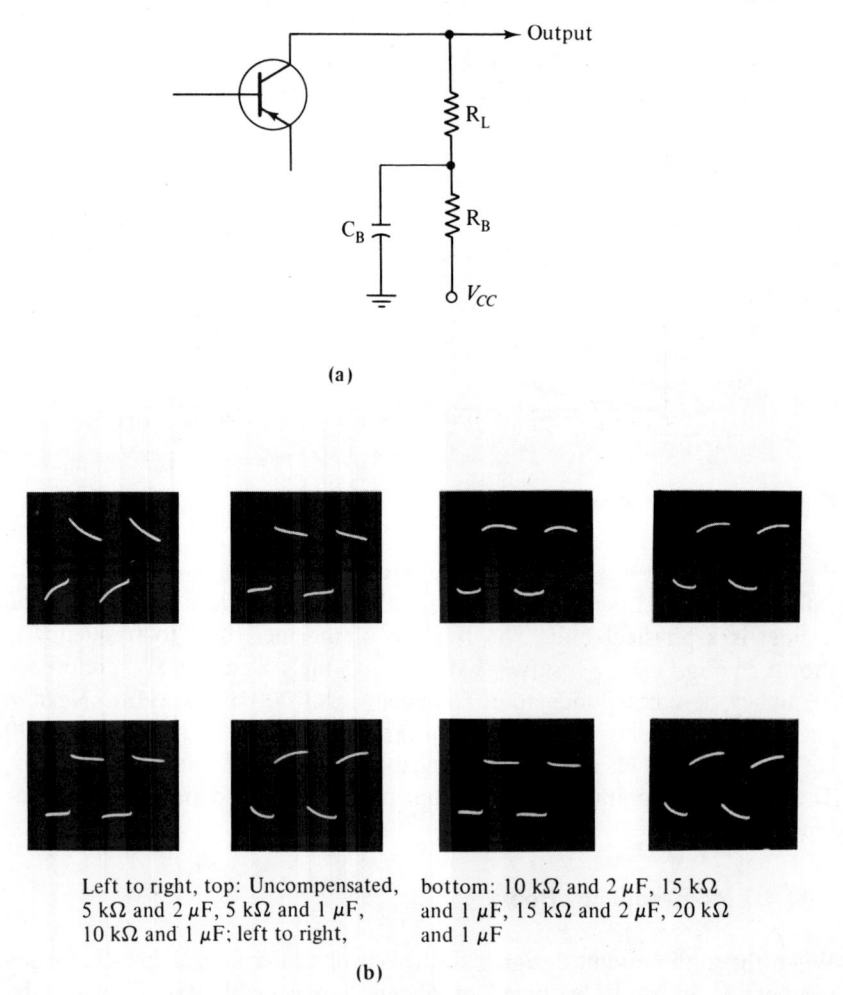

**Figure 1-34.** Low-frequency boost circuit: (a) $R_B$ and $C_B$ provide low-frequency boost; (b) examples of low-frequency square-wave response for various values of $R_B$ and $C_B$.

## 1-12 Presence Control

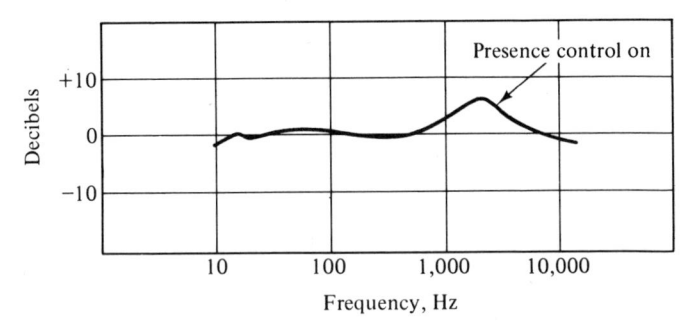

**Figure 1-35.** Effect of presence control action in a hi-fi system.

## 1-12 Presence Control

A presence control is employed in some high-fidelity systems to permit the listener to augment the output in the region of 2 kHz, as exemplified in Fig. 1-35. Listeners often prefer this frequency characteristic in comparison with a precisely flat midrange frequency response. A presence control is sometimes provided in the form of a level control for the midrange speaker in a speaker system. Alternatively, an RC filter network may be included in the preamplifier configuration. It is also possible to use a speaker crossover network that provides augmented response in the midrange.

chapter two

# FUNDAMENTALS OF AUDIO CIRCUIT DESIGN

## 2-1 Basic RC Circuitry

Audio circuit designers use RC circuitry in coupling and decoupling configurations, in tone-control and audio filter networks, in frequency-compensating circuits, and so on. Typical audio filter frequency characteristics are exemplified in Fig. 2-1; these are frequency-response curves for audio scratch and rumble filters. An RC filter is a frequency-selective configuration of resistors and capacitors, with or without active devices, that functions to attenuate or block current flow at some frequencies, with little opposition to current flow at other frequencies. This definition includes zero frequency (direct current). On the other hand, an RC frequency-compensating circuit, such as shown in Fig. 2-2, has an input–output relation that is independent of frequency. It is sometimes called an *all-pass filter* or all-pass network. It introduces attenuation without frequency discrimination to a reactive load.

A *passive filter* contains only passive components, such as resistance and capacitance. However, an *active filter* includes an active device, such as a transistor. RC filters can be grouped into high-pass, low-pass, band-pass, band-elimination (notch), and all-pass types. A low-pass filter is also called a high-cut filter; similarly, a high-pass filter is also termed a low-cut filter. High- and low-pass RC filters are commonly used in audio tone-control configurations, as exemplified in Fig. 2-3. Observe the voltage relations in a series RC circuit. The three circuit voltages combine to form a right triangle. This triangle can be circumscribed in a semicircle at any frequency. This graphical construction provides helpful visualization of tone-control action.

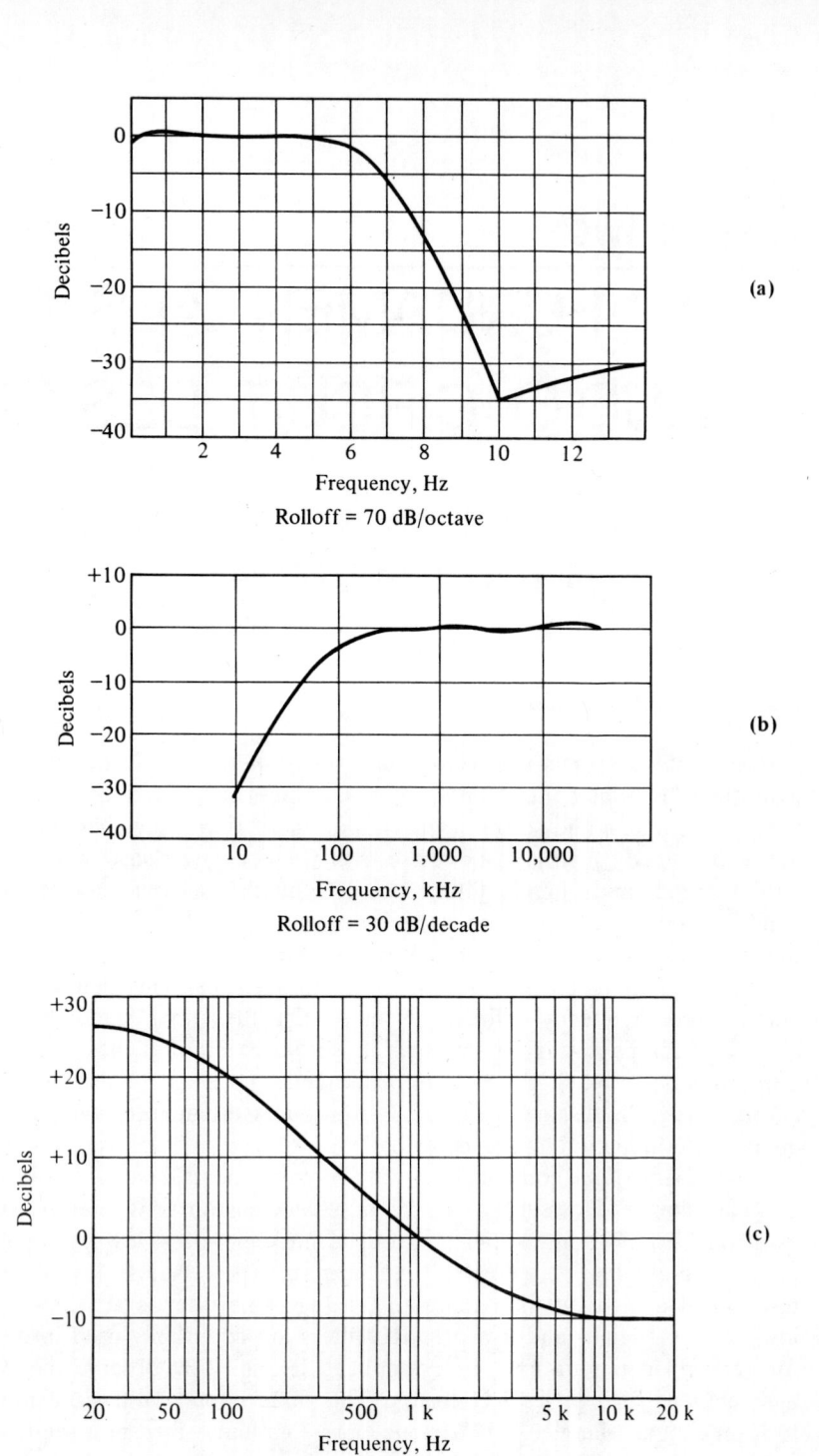

**Figure 2-1.** Typical audio filter characteristics: (a) scratch filter; (b) rumble filter; (c) NAB standard playback curve for 7.5 in./s.

## 2-1 Basic RC Circuitry

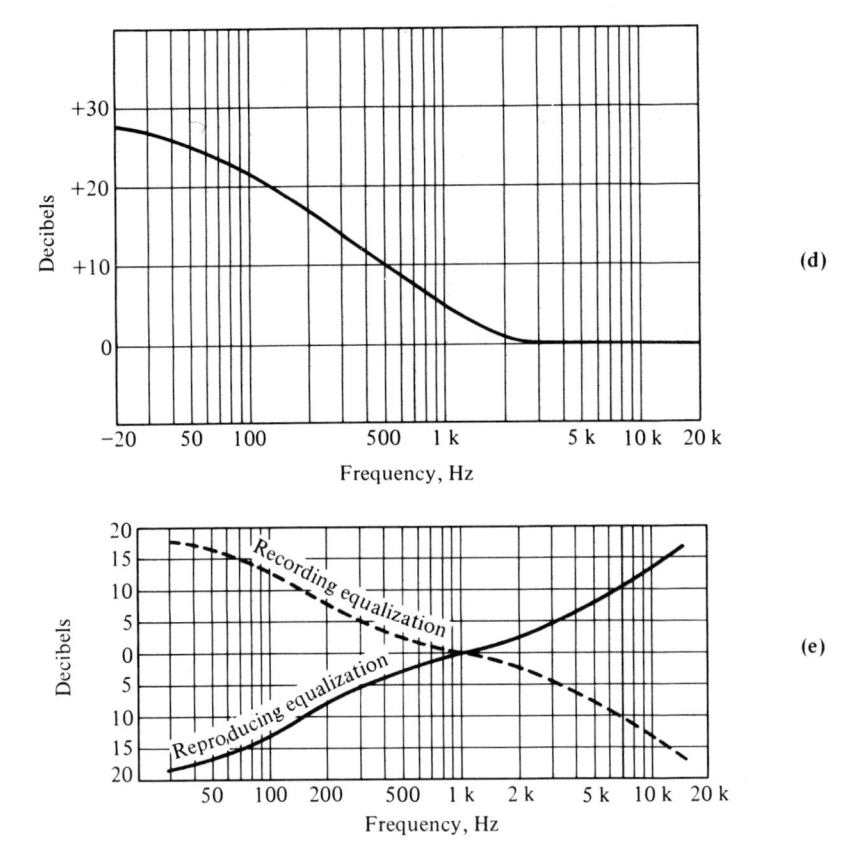

**Figure 2-1.** *(Continued)* (d) MRIA suggested playback curve for 3.75 in./s; (e) standard RIAA playback and recording equalization curves for disc recordings.

Note in Fig. 2-3a that point $P$ moves clockwise on the semicircle as the operating frequency increases. Conversely, point $P$ moves counterclockwise on the semicircle as the operating frequency decreases. The setting of a tone control establishes the reference frequency to which point $P$ corresponds. It is assumed that the output voltage in Fig. 2-3b is applied to a high-impedance load, such as a field-effect transistor. In other words, the output impedance of this tone control is comparatively high. If the tone control is used to energize a relatively low impedance load, such as a bipolar transistor, the characteristics of the tone control will be altered substantially, owing to circuit loading.

RC high-pass filter sections are also employed extensively as *coupling circuits*. These coupling circuits have an incidental filter function, in that they serve to block the flow of a dc component in a pulsat-

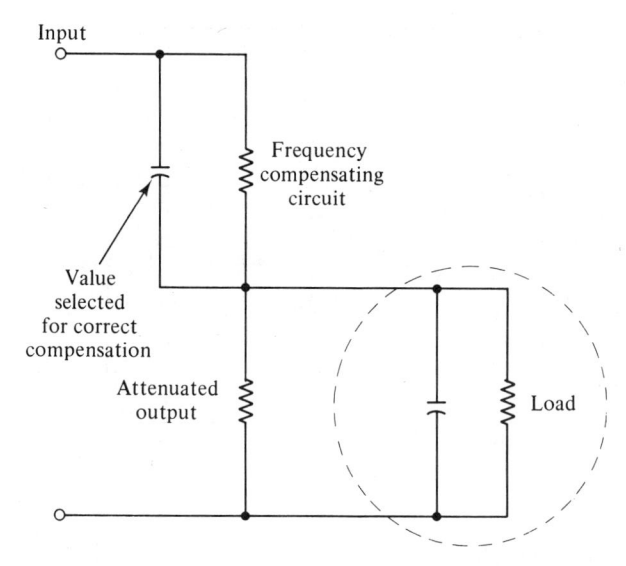

**Figure 2-2.** Example of an RC frequency-compensating circuit.

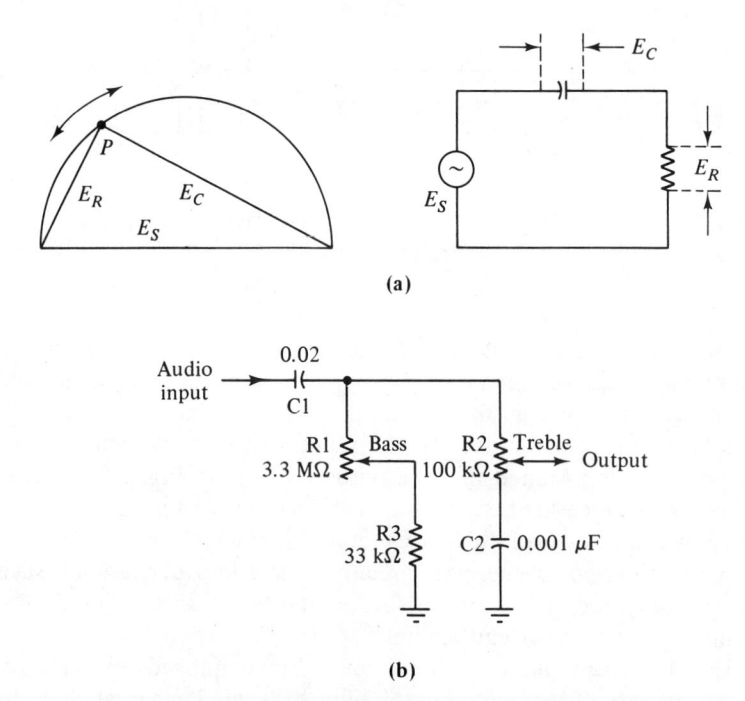

**Figure 2-3.** Basic audio tone-control configuration: (a) voltage relations in a basic RC circuit; (b) cascaded RC circuits in a simple tone-control network.

## 2-1 Basic RC Circuitry

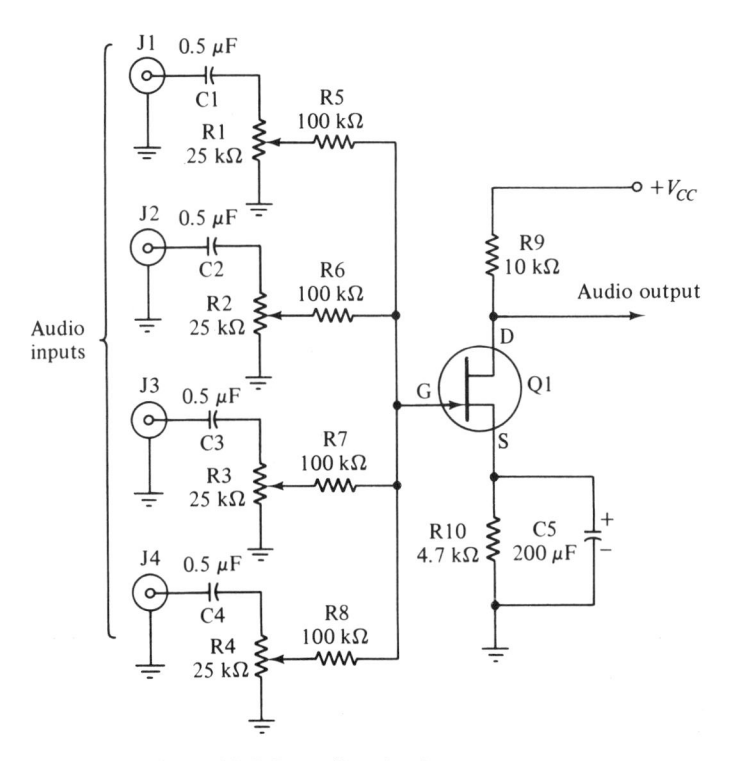

**Figure 2-4.** Audio mixer with RC coupling circuits.

ing-dc wave form, and to pass the ac component of the wave form with negligible frequency discrimination over the audio-frequency range. As an illustration, an audio mixer configuration with RC coupling circuitry is shown in Fig. 2-4. There is no serious *interaction* among the mixer controls, because the RC coupling circuits have a comparatively low internal impedance, and the 100-kΩ series resistors function as isolating resistors. It is practical for the circuit designer to use these 100-kΩ isolating resistors because the FET has an extremely high input impedance.

RC low-pass filter sections are widely utilized in *decoupling circuits*. A decoupling circuit comprises series resistance and shunt capacitance. It functions to greatly attenuate (bypass) the ac component of a pulsating-dc wave form and to pass the dc component of the waveform. Observe that the source circuit in Fig. 2-4 operates as a bypass arrangement. That is, it places terminal S at ac ground potential, whereas terminal S operates above dc ground potential. Thus decoupling and bypass circuits are closely related, although they do not serve identical functions. Note that R2 and C2 in Fig. 2-3 are often termed an RC filter section; similarly, R1, R3, and C1 are called an RC filter section. On the other hand, R1

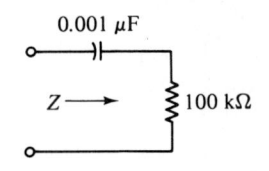

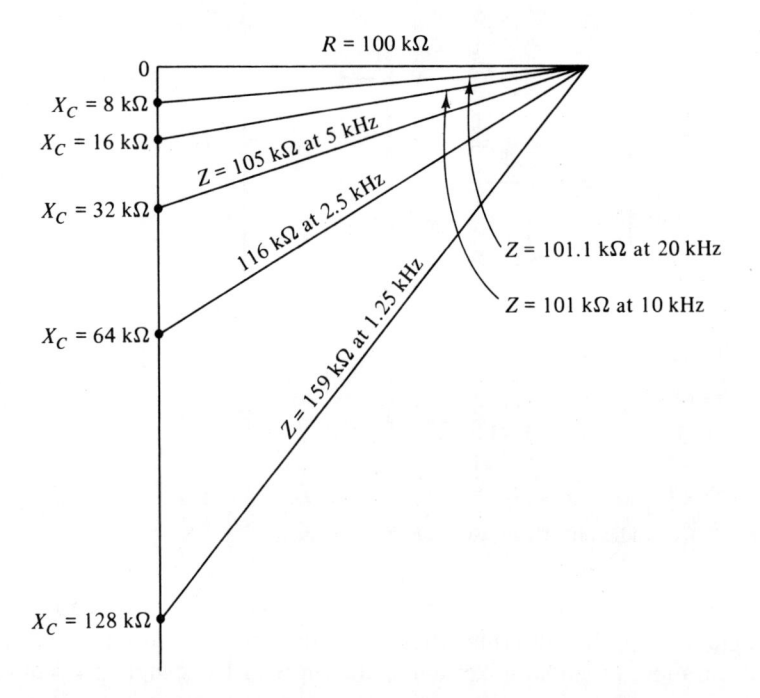

**Figure 2-5.** Impedance variational analysis of a series RC circuit.

and C1 in Fig. 2-4 are defined as an RC coupling circuit; R10 and C5 are defined as a bypass arrangement.

The impedance variation of a series RC circuit is diagrammed in **Fig. 2-5**. This is a simple example of variational analysis. Observe that the input impedance of the circuit changes rapidly from a low frequency to a moderately high frequency. Then the input impedance of the circuit changes less rapidly, and finally changes very slowly as the operating frequency varies from a moderate value to a high value. If we consider the voltage drop across the capacitor, the variational analysis is reversed with respect to frequency. Basic RC filters are defined as follows:

## 2-1 Basic RC Circuitry

1. A *low-pass filter* functions to permit current flow at all frequencies below a specified *cutoff frequency* with little (theoretically zero) loss, but to discriminate extensively (theoretically completely) against current flow at frequencies above the cutoff frequency.
2. A *high-pass filter* functions to permit current flow at all frequencies above a specified *cutoff frequency* with little (theoretically zero) loss, but to discriminate extensively (theoretically completely) against current flow at frequencies below the cutoff frequency.

The cutoff frequency of a filter is usually defined as the frequency on its frequency-response curve at which the output voltage is down -3 dB, or 70.7 percent, from its maximum value. Universal frequency-response charts for RC filter sections are shown in Fig. 2-6, with the -3-dB points noted. It is assumed that the RC sections work into a very high impedance load, such as the gate of an FET. If the load value is low, the frequency-response curves will be modified accordingly. Similarly, a reactive load modifies the frequency response of an RC filter. Consider the RC low-pass section depicted in Fig. 2-7a. If the section is driven by a voltage source, and operates into a resistive load $R_L$, as shown in Fig. 2-7b, then the equivalent circuit consists of $R$ and $R_L$ in parallel, as shown in Fig. 2-7c. In turn, the cutoff frequency becomes higher as a result of the resistive load.

Next, if the low-pass RC section works into a capacitor, it is evident that the equivalent circuit is calculated by simply adding the capacitance values of the section and the load. On the other hand, in case the low-pass RC section works into an RC load, the calculation of frequency response becomes more involved. In other words, a simplified equivalent circuit can be calculated for the configuration at any one chosen frequency. However, a simplified equivalent circuit does not exist for the configuration that holds true at all frequencies. In such a case, the circuit designer generally calculates the frequency response of the configuration at low, medium, and high frequencies. Then he draws an approximate frequency-response curve through these calculated points. If necessary, he will make additional calculations in the vicinity of the cutoff frequency, for example.

### Component Tolerances

In most situations, the cutoff frequency of an RC filter section is of basic concern. This value is affected by component tolerances as exemplified in

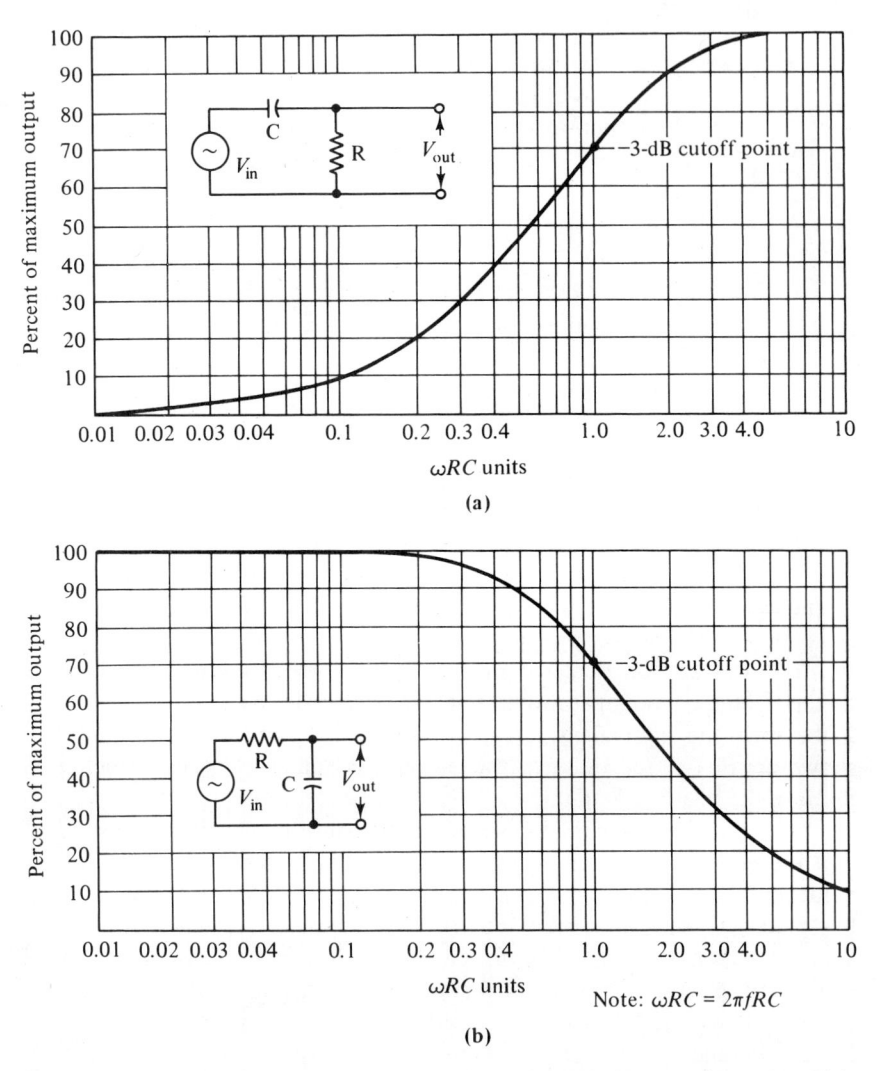

**Figure 2-6.** Universal frequency-response charts for RC filter sections: (a) high-pass section; (b) low-pass section.

Fig. 2-8. In this example, the tolerance on $R$ and $C$ values is ±20 percent. The cutoff frequency is equal to $1/(2\pi RC)$. Accordingly, if $R$ has a bogie value of 10 kΩ and $C$ a bogie value of 0.001 μF, the -3-dB point on the frequency-response curve will occur at 15,920 Hz. This is the design-center, or bogie cutoff frequency. When ±20 percent component tolerances are taken into consideration, the cutoff-frequency limits are found to be 24,875 and 11,055 Hz. Thus the tolerance on the cutoff frequency of the RC low-pass section can be expressed as 15,920 + 8,855 − 4,865

## 2-1 Basic RC Circuitry

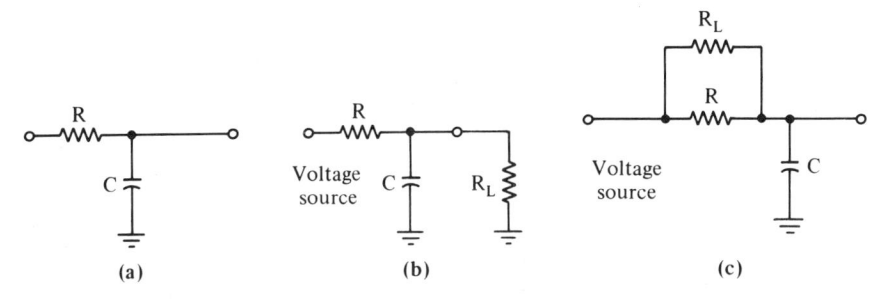

**Figure 2-7.** Example of an equivalent circuit for a loaded RC low-pass section: (a) $Z_{in} = R - 1/j\omega C$ (b) section with resistive load; (c) equivalent circuit for (b), $Z = \dfrac{RR_L}{R + R_L} - 1/j\omega C$.

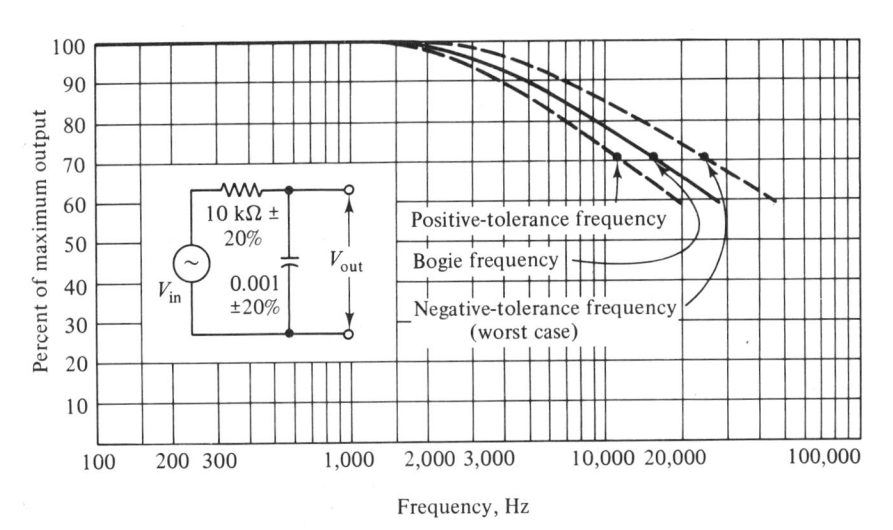

**Figure 2-8.** Example of ±20 percent component tolerances on cutoff frequency.

Hz. The worst-case condition (maximum deviation) occurs when both the $R$ and $C$ tolerances have their negative limit values.

Note that, in the foregoing example, if the tolerance on $R$ is at its positive limit and the tolerance on $C$ is at its negative limit, these tolerances tend to cancel, although they do not cancel completely. In this situation, the cutoff frequency becomes 16,583 Hz, which is 668 Hz greater than the bogie value of 15,920 Hz. In summary, the cutoff-frequency errors that result from ±20 percent tolerances on the $R$ and $C$ values are as follows:

1. If $C$ has its bogie value and the value of $R$ is 20 percent low, the cutoff-frequency error is 25 percent. Or if $R$ has its bogie value and the value of $C$ is 20 percent high, the cutoff-frequency error is 25 percent.
2. If $C$ has its bogie value and the value of $R$ is 20 percent high, the cutoff-frequency error is approximately 17 percent. Or if $R$ has its bogie value and the value of $C$ is 20 percent high, the cutoff-frequency error is approximately 17 percent.
3. If the value of $R$ is 20 percent low and the value of $C$ is also 20 percent low, the cutoff-frequency error is 56 percent (worst case).
4. If the value of $R$ is 20 percent high and the value of $C$ is also 20 percent high, the cutoff-frequency error is 31 percent.
5. If the value of $R$ is 20 percent high and the value of $C$ is 20 percent low, the cutoff-frequency error is 4 percent.

## 2-2 Input Impedance of an RC Section

Consider the effect of component tolerances on the input impedance of an RC section. It will be assumed, for the purpose of basic analysis, that the load value is sufficiently high that it can be neglected. In effect, the output terminals of the RC section are considered to be open circuited. In turn, the input impedance of either a high-pass or low-pass RC section is infinite at zero frequency (direct current), and is equal to $R$ at infinite frequency. With reference to Fig. 2-8, the input impedance of the RC section is 14,142 $\Omega$ at the cutoff frequency of 15,920 Hz, as indicated in Fig. 2-9. At the -3-dB cutoff frequency, the reactance of $C$ is $-j10,000$ $\Omega$, which combines in quadrature with the 10,000-$\Omega$ resistance to form an impedance of 14,142 $\Omega$ with a phase angle of 45 degrees.

Next, the effect of component tolerances on this input-impedance value is as follows. If the resistor has an actual value of 12,000 $\Omega$, and the capacitor has an actual value of 0.0012 $\mu$F, the input impedance of the RC section at its cutoff frequency becomes 17,000 $\Omega$; compare this value with the bogie value of 14,142 $\Omega$. This is a tolerance error of 20 percent. Next, if the resistor has an actual value of 8,000 $\Omega$, and the capacitor has an actual value of 0.0008 $\mu$F, the input impedance of the RC section becomes 11,300 $\Omega$, as compared with a bogie value of 14,142 $\Omega$. This is a tolerance error of 28 percent, and it represents the worst-case value in this example. In summary, the input impedance of the RC section could range from 11,300 to 17,000 $\Omega$ when the components have $\pm$ 20 percent tolerances.

## 2-3 Output Impedance of an RC Section

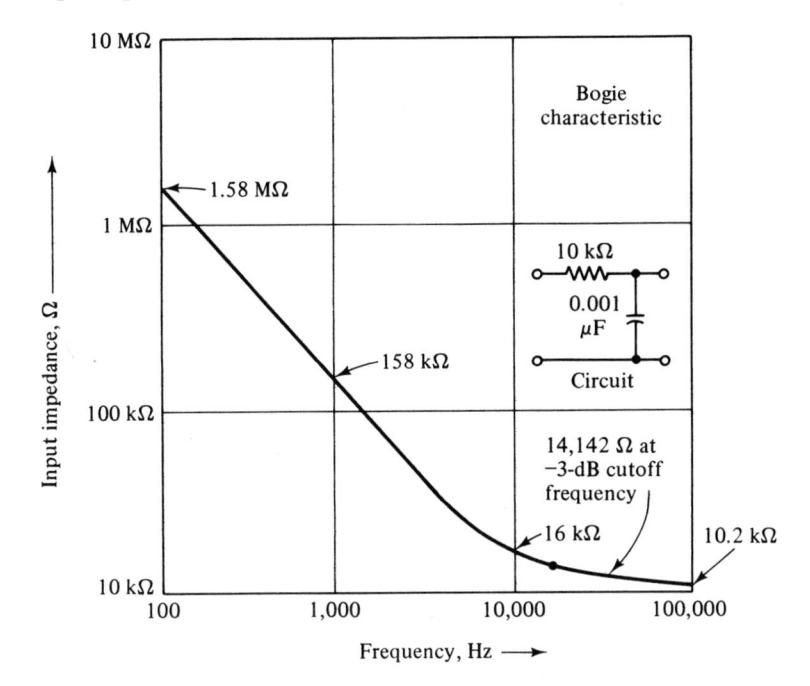

**Figure 2-9.** Input-impedance variation of an RC high-pass filter section versus frequency.

## 2-3 Output Impedance of an RC Section

Consider next the effect of component tolerances on the output impedance of an RC section. With reference to Fig. 2-6, the output impedance of either section is defined as the impedance "looking backward" from the output terminals into the circuit; a constant-voltage source will be assumed, or we will consider that the input terminals are short circuited. In turn, the output impedance of either the high- or low-pass section consists of a parallel combination of resistance and capacitance. With reference to Fig. 2-8, the section has an output impedance of 10,000 Ω at zero frequency (direct current) and zero output impedance at infinite frequency. At the cutoff frequency of 15,920 Hz, the 10-kΩ resistor operates in parallel with $-j10,000$ Ω of capacitive reactance. Accordingly, the output impedance at the cutoff frequency is 7,072 Ω, as shown in Fig. 2-10. This output impedance is capacitive, and its phase angle is 45 degrees at the -3-dB cutoff frequency.

Next consider the output impedance variation versus frequency of an RC section, as depicted in Fig. 2-11. This construction is somewhat

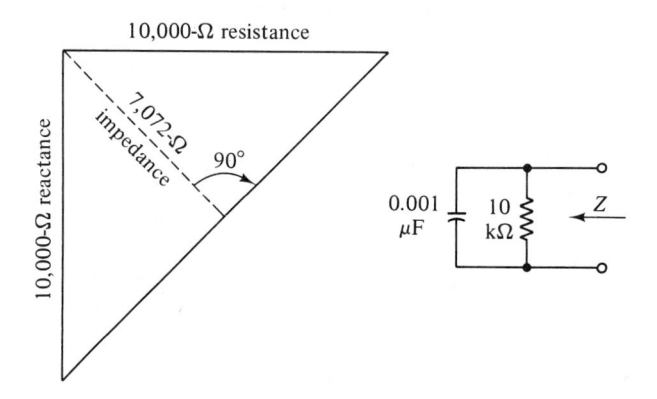

**Figure 2-10.** Impedance value of parallel resistance and capacitance.

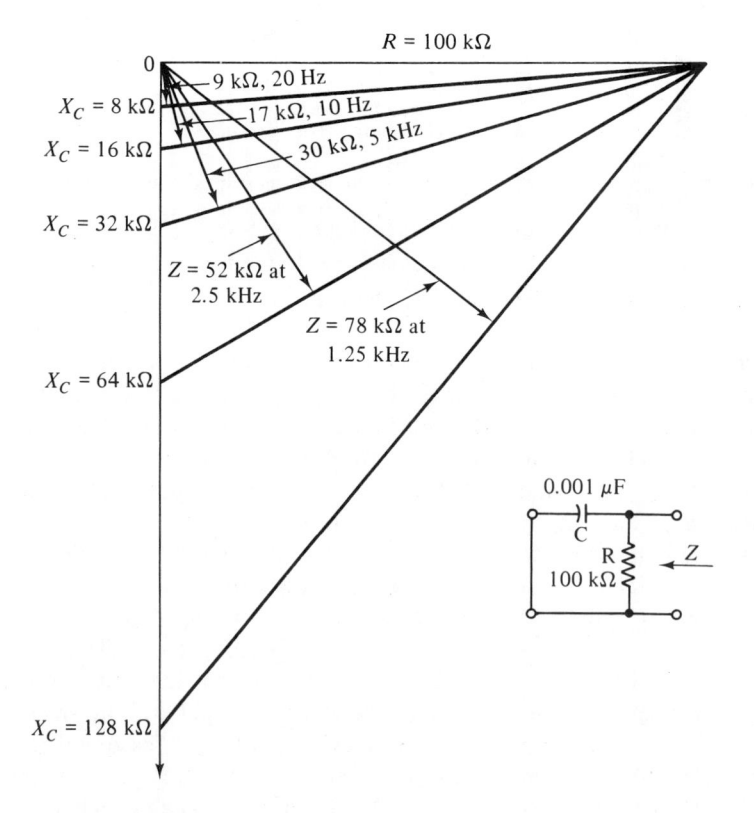

**Figure 2-11.** Example of output impedance variation versus frequency for an RC section.

## 2-3 Output Impedance of an RC Section

similar to that shown in Fig. 2-5, except that we now consider the lengths of the perpendiculars dropped from the origin to the various series-impedance vectors (compare with the construction shown in Fig. 2-10). It will be observed that the impedance variation depicted in Fig. 2-11 is rapid at low frequencies, and that it becomes progressively slower at higher frequencies of operation. This variation is clarified by the construction in Fig. 2-12; it shows that the impedance of a parallel RC section versus frequency follows a semicircular locus.

Since the output voltage is taken across the resistor in a high-pass RC filter section, the output voltage leads the input voltage. Conversely, since the output voltage is taken across the capacitor in a low-pass RC filter section, the output voltage lags the input voltage.

Consider next the effect of tolerances on the output-impedance value of an RC filter section. If the resistor has an actual value of 12,000

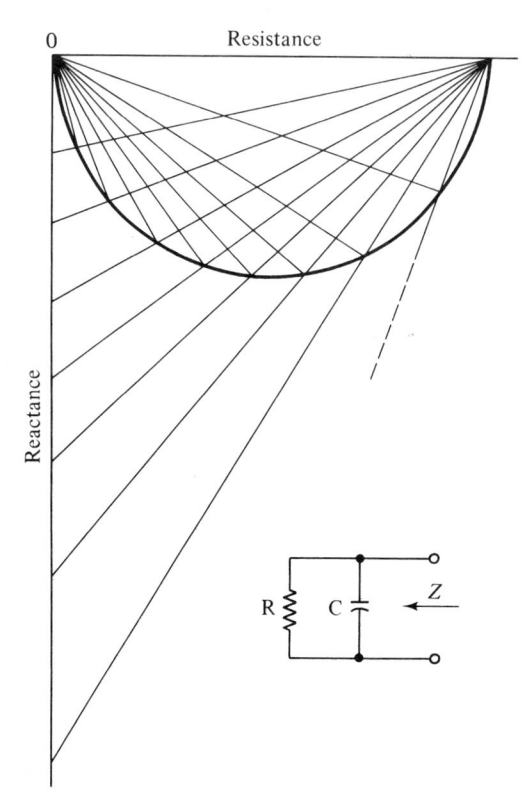

**Figure 2-12.** Impedance variation of a parallel RC section follows a semicircular locus.

Ω, and the capacitor has an actual value of 0.012 μF, the output impedance of the RC filter section at its cutoff frequency becomes 8,500 Ω, as compared with a bogie value of 7,072 Ω. This is a tolerance error of 20 percent. Next, if the resistor has an actual value of 8,000 Ω, and the capacitor has an actual value of 0.0008 μF, the output impedance of the RC filter section becomes 5,650 Ω, as compared with a bogie value of 7,072 Ω. This is a tolerance error of 20 percent. Accordingly, both the high- and low-tolerance limits represent worst-case conditions. In summary, the output impedance of the RC filter section in this example could range from 5,650 to 8,500 ohms.

## Insertion Loss

An RC filter section dissipates energy; this dissipation is called its *insertion loss*. Insertion loss is defined as the difference between the power received at the load before and after the insertion of the filter section between the source and the load. It is understood that the measurement between the two conditions of operation is made within the pass band of the filter. With reference to Fig. 2-13a, the 1-kΩ load dissipates 1 milliwatt (mW) of power from the 1-V generator at any frequency. Because a low-pass RC filter section will be considered, circuit operation will be restricted to a low audio frequency. Next, in Fig. 2-13b, a low-pass RC

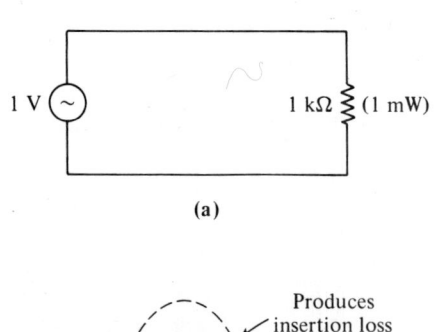

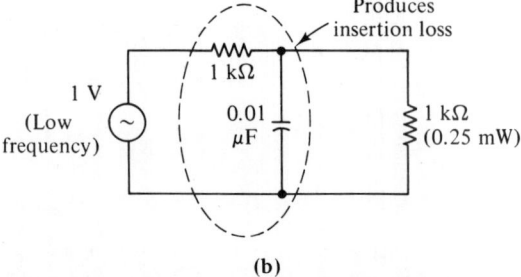

**Figure 2-13.** Insertion loss of the RC section is 0.75 mW.

## 2-4 Filter Output Voltage

filter section comprising a 1-kΩ series resistor and a 0.01-μF shunt capacitor has been inserted. In turn, the power dissipated by the 1-kΩ load drops to 0.25 mW. Therefore, the insertion loss of the filter section is equal to 0.75 mW.

## 2-4 Filter Output Voltage

With reference to the high- and low-pass RC filter sections shown in Fig. 2-6, it is instructive to observe their *rolloff characteristics*. Rolloff is defined as the manner in which the amplitude-frequency characteristic of a filter varies as it approaches its frequency limits. The output voltage from a high-pass filter is less than its input voltage, although the output amplitude approaches the input amplitude at very high operating frequencies. Conversely, the output voltage level from the low-pass filter is less than its input voltage level, although the output voltage approaches the input voltage at very low operating frequencies. At any frequency, the ratio of output voltage to input voltage for a high-pass section is equal to $R/Z$, and for a low-pass section it is equal to $X/Z$. Ratio values are easily determined from the charts in Fig. 2-6.

Rolloff is generally expressed in terms of minus decibels per octave or of minus decibels per decade. Refer to Fig. 2-14. This is a basic example of rolloff. An *octave* is defined as an interval over which the operating frequency doubles. A *decade* is defined as an interval over which the operating frequency increases 10 times. In this example, the output level decreases by 3.6 dB when the operating frequency is doubled. Otherwise stated, the rolloff is equal to $-3.6$ dB/octave. This is an example of uniform rolloff. Observe that the output level decreases by 12 dB when the operating frequency increases 10 times. Therefore, the rolloff is equal to $-12$ dB/decade. In the case of nonuniform rolloff, it is sometimes desired to determine the rate of rolloff at a point, as exemplified in Fig. 2-15. The *rate of rolloff* is defined as the slope of the tangent to the frequency-response curve at the given point. In this example, the rate of rolloff is 6 dB/octave.

With reference to Fig. 2-6, it is observed that the output voltage for an RC filter section has an input–output ratio of 1.57 through the $-3$-dB cutoff point. This corresponds to a rolloff rate of approximately $-3.9$dB/octave. As in many practical situations, this is a case of nonuniform rolloff. In other words, if we consider the input–output voltage ratio over the 0.1 to $-1$ or over the 1.0 to 10 decades for the high- and low-pass sections, respectively, this ratio is seen to be 7.07, corresponding to $-17$ dB/decade. This figure is only an approximation, because the rolloff is not linear over the decade interval. In practical design work, such approximations are often adequate from an engineering viewpoint.

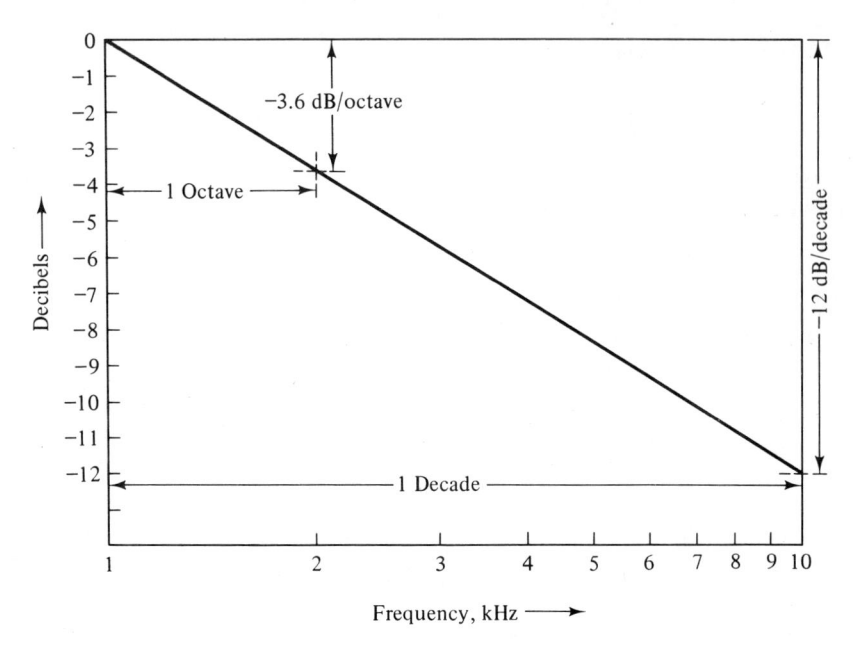

**Figure 2-14.** Example of decibel rolloff per octave and per decade.

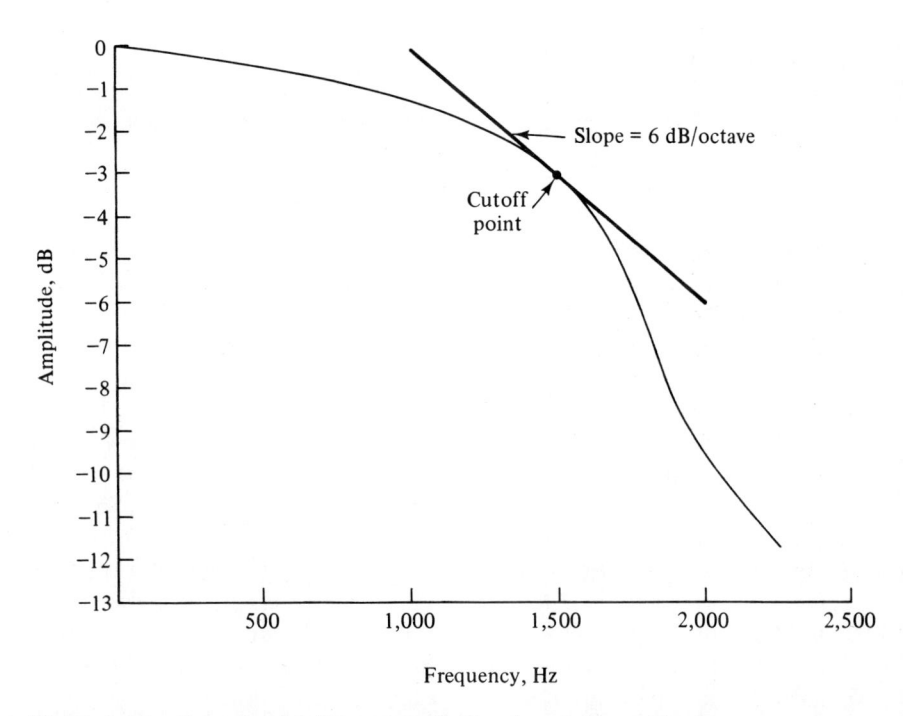

**Figure 2-15.** Example of 6 dB/octave rolloff at the −3-dB cutoff point.

## 2-5 Frequency Response of Cascaded RC Sections

A common cascaded RC configuration is depicted in Fig. 2-16. For example, successive RC coupling circuits are employed in basic amplifier arrangements. In a preliminary analysis of network response, it is assumed that the device has a gain of unity, and that its only function is to

(a)

(b)

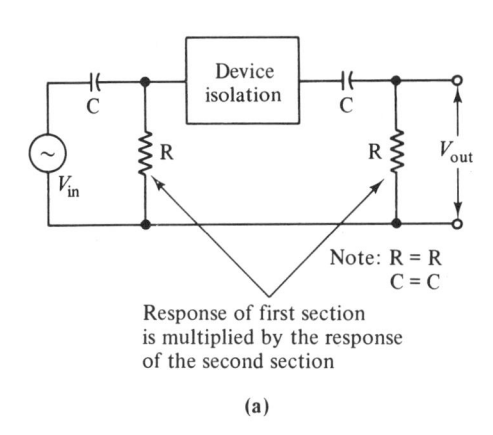

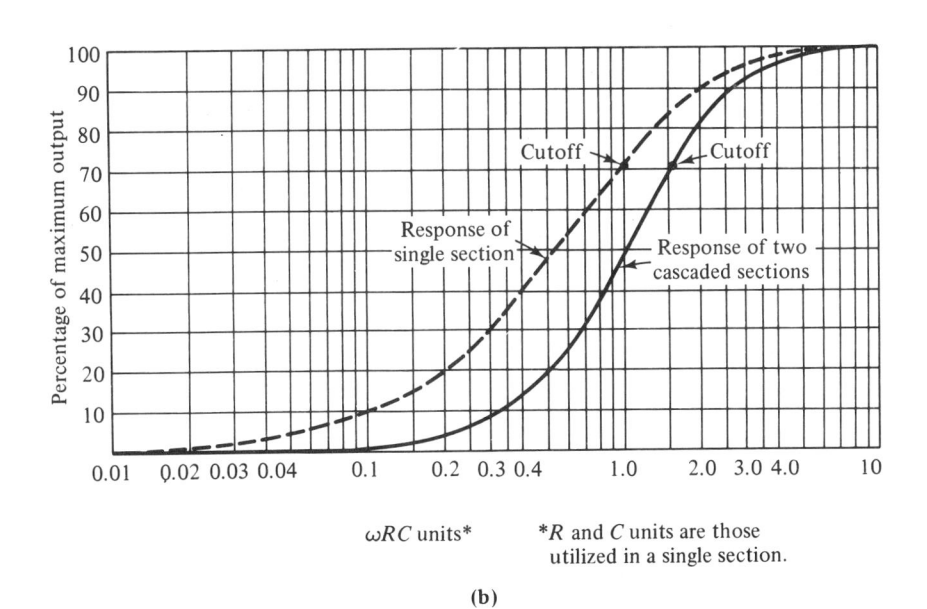

**Figure 2-16.** Two cascaded RC high-pass sections with device isolation: (a) configuration; (b) universal frequency-response chart.

provide isolation between the first and second RC sections. This device might be a field-effect transistor, for example, biased for unity gain. Unless isolation is provided between the two RC sections, the first section will be loaded by the input impedance of the second section, and the network response will not be the same. With device isolation included, the frequency response of the first section is shown by the dotted curve in Fig. 2-16b; the second section has the same frequency response. However, when the second section is driven by the output from the first section, a different frequency response results, as shown by the solid curve in Fig. 2-16b.

We observe that the cascaded configuration has a steeper rolloff than a single RC section. In other words, the rolloff through the $-3$-dB cutoff point is 5 dB/octave for the two-section arrangement, whereas the rolloff through the $-3$-dB cutoff point is 3.9 dB/octave for a single section. Moreover, the cutoff frequency is higher for the cascaded configuration than for a single section. That is, the $-3$-dB cutoff frequency for the cascaded sections corresponds to $\omega RC = 1.55$, approximately, compared with $\omega RC = 1$ for a single RC section. This shift in cutoff frequency for the cascaded configuration is an aspect of its comparatively steep rolloff. It is apparent that a three-section cascaded arrangement would have a still higher $-3$-dB cutoff frequency.

The two-section cascaded configuration has an input–output relation wherein the output amplitude is equal to the square of the input amplitude. Thus the output level from the first section is 0.707 of the input level at the frequency of section cutoff ($\omega RC = 1$). Next, the second section has an output level that is equal to 0.707 of its input level. Accordingly, the output level from the second section is equal to 50 percent of the input level to the first section, at $\omega RC = 1$.

Next consider the effect of component tolerances on the cutoff frequency of a two-section cascaded RC configuration that includes device isolation. Because of the comparatively steep rolloff of the two-section configuration, component tolerances have a lesser effect on cutoff-frequency shift than was noted for a one-section arrangement. Refer to Fig 2-8. Consider the $-3$-dB cutoff frequencies that will occur when this RC section is cascaded with a similar RC section (device isloation included). It is apparent that the two-section cutoff frequency will occur at the 84.1 percent level with respect to the single-section characteristic. In turn, the bogie cutoff frequency for two sections in cascade is approximately 7,500 Hz. If the $R$ and $C$ components have $\pm 20$ percent tolerances, the cutoff frequencies become approximately 5,800 Hz for positive tolerances and approximately 11,000 Hz for negative tolerances. This latter limit represents the worst-case condition in most applications.

## 2-6 RC Bandpass Filter Characteristics

In theory, an ideal bandpass filter has a characteristic such as shown in Fig. 2-17. However, in practice, the rolloff is not infinite, and the top of the frequency characteristic is not perfectly flat topped. An RC bandpass filter is formed by connecting a high-pass section in series with a low-pass section. A simple example of an RC bandpass filter configuration is shown in Fig. 2-18. Component values are chosen to obtain a $-3$-dB low-frequency cutoff of 100 Hz and a high-frequency cutoff of 10 kHz, approximately. The low-end and high-end rolloff and the cutoff frequencies follow directly from the relations depicted in Fig. 2-6. From a design viewpoint, tolerances on component values in this bandpass filter arrangement have the same effect on frequency response as in their low- and high-pass counterparts. Observe that there is comparatively little interaction between the high-pass and low-pass sections in the example of Fig. 2-18, because the ratio of the cutoff frequencies is 100 to 1. On the other hand, if the audio-circuit designer chooses a cutoff-frequency ratio of 10 to 1 or less, circuit interaction must be taken into account.

When the cutoff-frequency ratio of an RC bandpass filter such as depicted in Fig. 2-18a is chosen at 10 to 1 or less, the designer can no longer analyze its characteristics accurately by individual consideration of its low- and high-pass sections. Instead, he must now treat the bandpass filter network as a series–parallel configuration over its entire pass band.

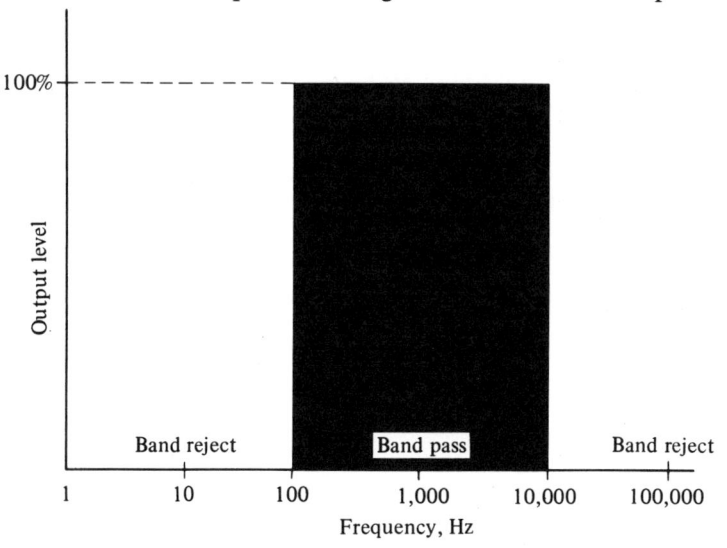

**Figure 2-17.** Theoretically ideal bandpass filter characteristic.

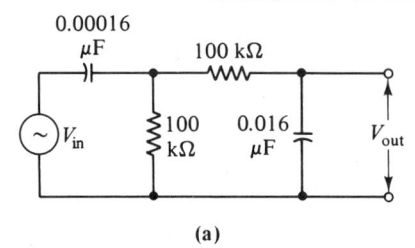

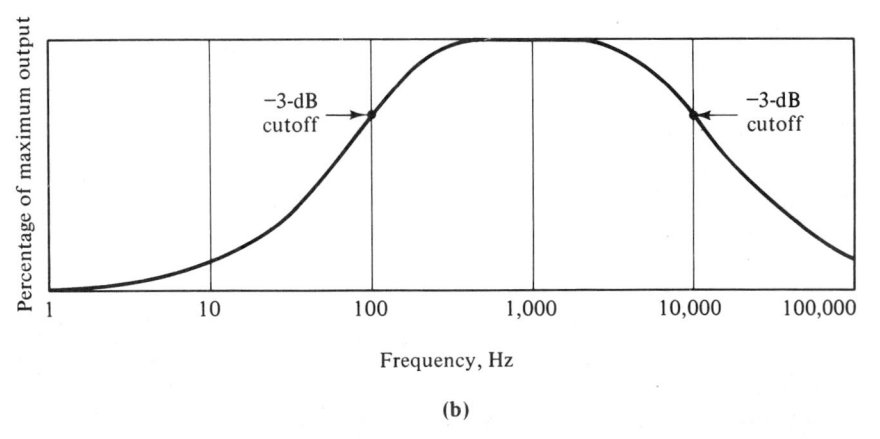

**Figure 2-18.** Example of simple RC bandpass filter arrangement: (a) configuration; (b) frequency-response curve.

Consequently, the response calculations become comparatively involved. On the other hand, the response of the bandpass filter configuration can be easily determined from the charts in Fig. 2-6, if device isolation is included between the high- and low-pass sections, as shown in Fig. 2-19. It is convenient to assume a device gain of unity, so that the universal-chart data are directly applicable. Note that if gain is provided by the device the frequency response remains unchanged, but the gain factor must be carried through the calculations from the chart data.

If comparatively rapid rolloff is desired in the frequency response of an RC bandpass filter, multisection high- and low-pass configurations can be employed, as exemplified in Fig. 2-20. Note that both the input and output networks must be analyzed as series–parallel configurations, because of circuit loading. However, there is no interaction between the input and output networks. If one RC section has a cutoff frequency of 1,900 Hz, and if it is directly connected to a similar RC section, the cutoff frequency for the two-section arrangement becomes 800 Hz. This is a frequency shift of approximately 57 percent, compared with a frequency shift of 45 percent when device isolation is employed between two similar sections.

## 2-7 RC Band-Elimination (Notch) Filter

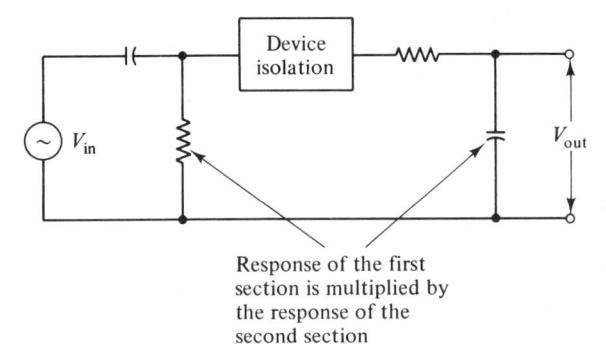

**Figure 2-19.** Basic RC bandpass filter arrangement with device isolation.

(Response of the first section is multiplied by the response of the second section)

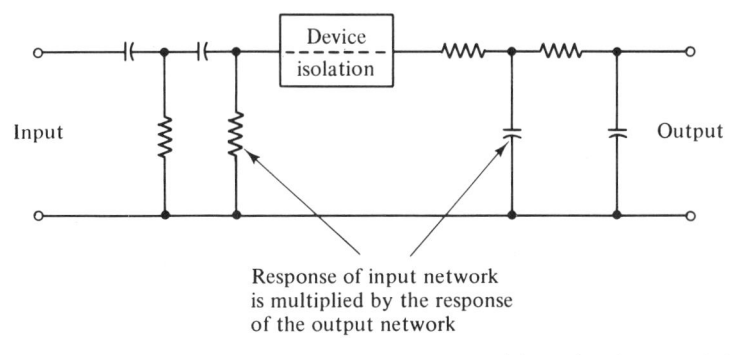

**Figure 2-20.** Bandpass RC filter arrangement with double-section input and double-section output.

(Response of input network is multiplied by the response of the output network)

## 2-7  RC Band-Elimination (Notch) Filter

A theoretically ideal bandpass or notch filter frequency-response characteristic is pictured in Fig. 2-21. In practice, however, the rolloff is not infinite, and the dip has a rounded minimum contour. A widely used type of filter configuration is the RC parallel-T network, shown in Fig. 2-22. This is essentially a low-pass RC tee section connected in parallel with a high-pass RC tee section. As noted in the diagram, a basic RC parallel-T network employs component values such that a null (zero output) occurs at a specified frequency. Null action results from equal output amplitudes from the two RC sections, with a 180-degree phase difference between the two outputs. The shunt resistor has one half the value of either of the series resistors, and the shunt capacitor has double the value of either of the series capacitors. When close-tolerance components are utilized, a zero-

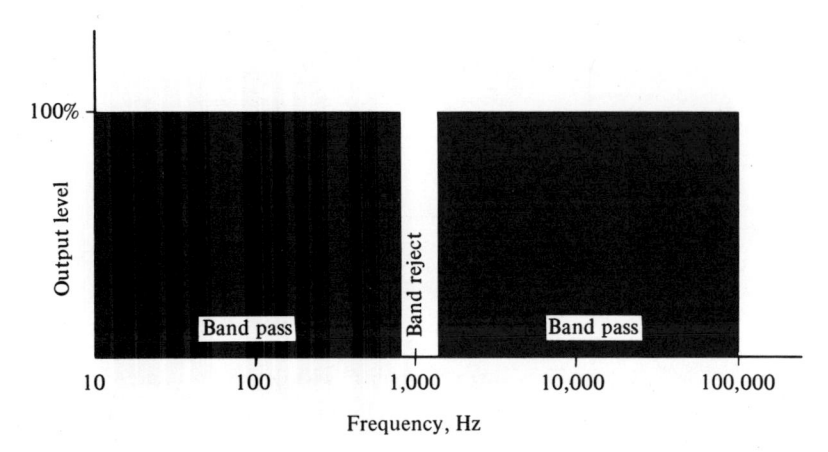

**Figure 2-21.** Theoretically ideal response of a band-reject (notch) filter.

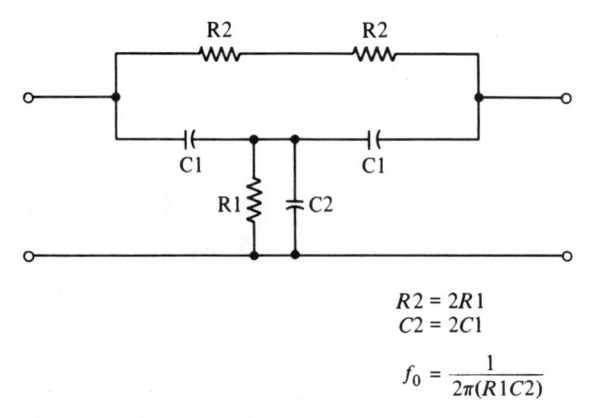

$$R2 = 2R1$$
$$C2 = 2C1$$
$$f_0 = \frac{1}{2\pi(R1C2)}$$

**Figure 2-22.** Basic parallel-Tee RC band-elimination filter.

voltage null output is obtained at frequency $f_0$. It is assumed that a true sine-wave voltage is applied to the network. That is, if the applied wave form contains harmonics, a shallow (incomplete) null will result at frequency $f_0$.

A frequency-response curve for a parallel-T RC filter section is exemplified in Fig. 2-23. This configuration provides a null frequency of 4.4 kHz, approximately. Its −3-dB bandwidth is about 4.9 kHz. Component tolerances have a considerable effect on the null frequency. For example, when all tolerances are at their 20 percent low limit, the null frequency becomes 56 percent higher than bogie. This is the worst-case condition. When all tolerances are at their 20 percent high limit, the null frequency becomes 31 percent lower than bogie. Note that a complete null

## 2-8 Active High/Low-Pass RC Filter Network

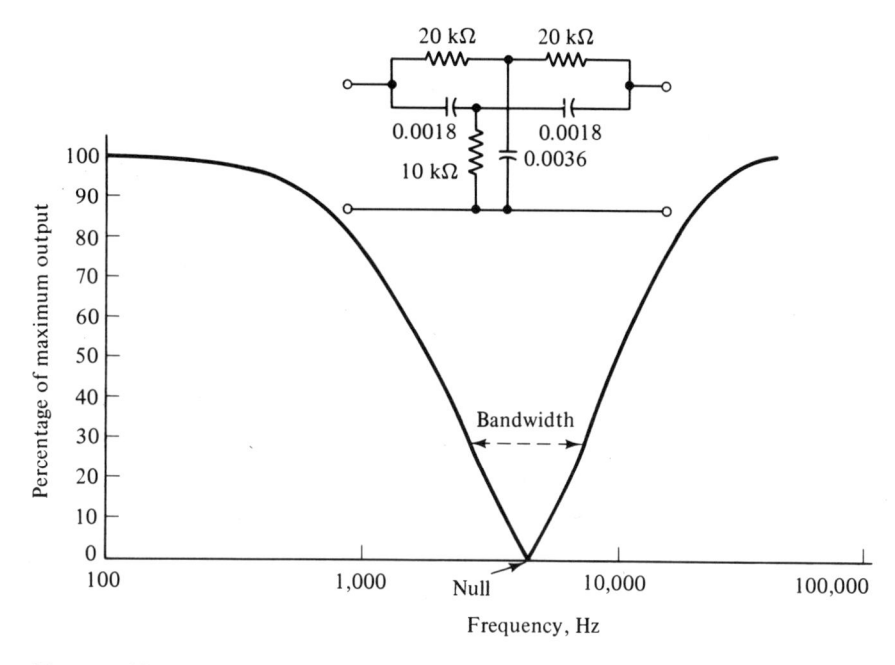

**Figure 2-23.** Example of a frequency-response curve for a parallel-Tee RC filter section.

is obtained when all tolerances are 20 percent high or 20 percent low, because the R1 to R2 and C1 to C2 ratios remain the same. On the other hand, if some of the component tolerances are high and others are low, a complete null is not obtained.

## 2-8 Active High/Low-Pass RC Filter Network

An example of an active high/low-pass RC filter network is shown in Fig. 2-24. This is a bass–treble tone-control arrangement that provides a bass cut–boost section and a treble cut–boost section. Equivalent circuits for the extreme settings of the controls are shown in the diagram. Transistors Q1 and Q2 provide gain and overcome the insertion loss of the RC network. In addition, Q2 provides isolation between the bass-control and treble-control sections. This isolation effectively eliminates interaction between the bass and treble control settings. The effect of tone-control settings on the frequency response of the network is shown in Fig. 2-25. A total range of approximately ±15 dB is provided.

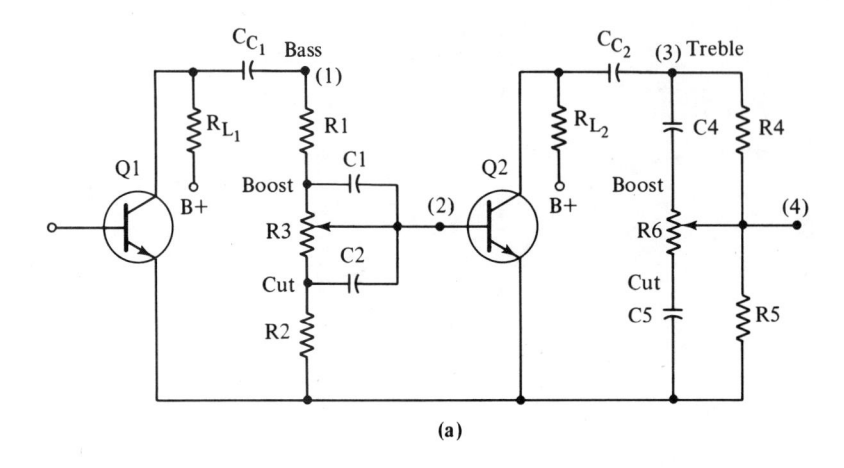

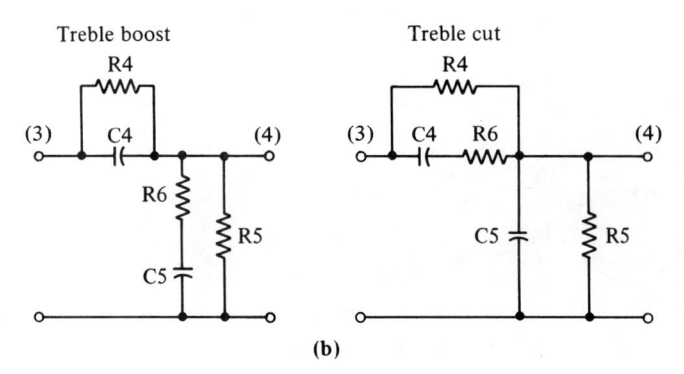

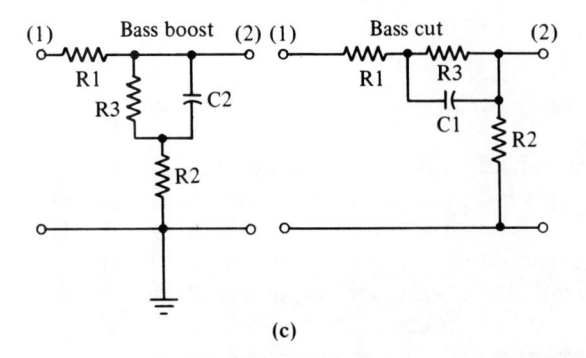

**Figure 2-24.** Typical configuration for an active bass–treble tone control: (a) bass–treble RC tone-control sections with device isolation; (b) treble section equivalent circuits at extreme control settings; (c) bass section equivalent circuits at extreme control settings.

## 2-8 Active High/Low-Pass RC Filter Network

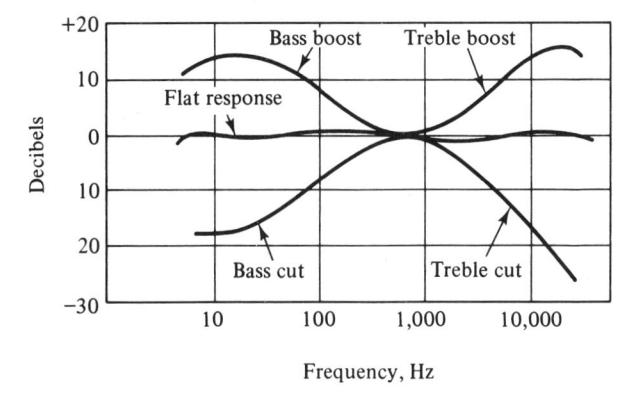

**Figure 2-25.** Frequency-response variations provided by a tone control.

chapter three

# PRINCIPLES OF AUDIO AMPLIFIER DESIGN

## 3-1 General Considerations

An audio amplifier in a high-fidelity system has the basic plan depicted in Fig. 3-1. That is, it comprises a preamplifier section and a power-amplifier section. The audio-amplifier channel in a typical hi-fi system has a maximum usable gain (MUG) of more than 5,000 times, as shown in Fig. 3-2. Although industry standards have not been established, it is generally agreed that high-fidelity reproduction denotes a frequency response from at least 20 Hz to 20 kHz, with a *total harmonic distortion*

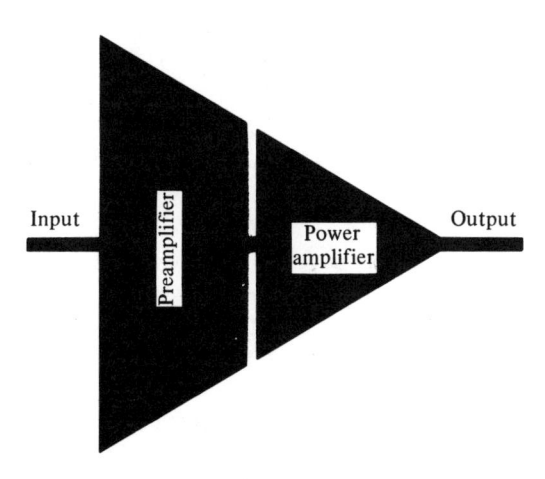

**Figure 3-1.** Basic plan of a high-fidelity audio amplifier.

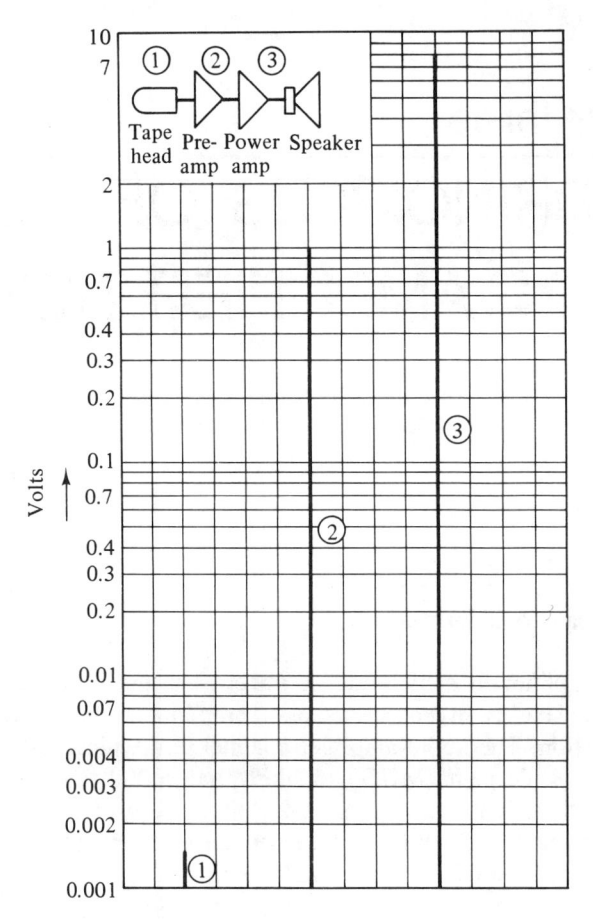

**Figure 3-2.** Signal-voltage levels at amplifier inputs and outputs in a high-fidelity system.

of less than 1 percent at maximum rated power output. This basic distortion measurement is made at a frequency of 1 kHz. *Intermodulation distortion* is comparable in a general way to harmonic distortion; however, the former is measured with a two-tone test signal, whereas the latter is measured with a single-frequency signal. A more sophisticated type of distortion measurement, called *transient intermodulation distortion,* is also recognized as a basic criterion of high-fidelity performance. These topics are detailed subsequently. Table 3-1 lists typical audio-amplifier specifications.

*Power bandwidth* is defined as the frequency range between an upper and a lower limit, at a power level 3 dB below maximum rated power output, where the harmonic distortion starts to exceed the value that oc-

## 3-2 Fundamental Amplifier Design Procedures

**Table 3-1. Typical Audio-Amplifier Specifications**

*Input Sensitivity:* Phono, 2 mV.
*Hi-level* (tuner, tape, aux., tape mon, and tape dub), 180 mV.
*Input Overload:* Phono, over 100 mV. Hi-level, over 10 V.
*Hum and Noise:* Phono, −65 dB (2-mV input); hi-level, −85 dB (180-mV input).
*Input Impedance:* Phono, 47 k$\Omega$. Hi-level, 50 k$\Omega$.
*Frequency Response:* Phono, ±0.5 dB, 30 Hz to 15 kHz. Hi-level, ±0.2 dB, 20 Hz to 20 kHz.
*Rated Output:* 1.5 V at rated sensitivity.
*Output Impedance:* 500 $\Omega$.
*Tape Output:* 180 mV.
*Harmonic and IM Distortion:* Less than 0.05% from 20 Hz to 20 kHz at 1.5-V output.
*Channel Separation:* 50 dB at 1 kHz.
*Filters:* Low, −3 dB at 15 Hz, −12 dB octave/slope. High, −3 dB at 7 kHz, −12 dB octave/slope.
*Headphone Amplifier, Harmonic Distortion:* 0.1% (4 V min. into 100 $\Omega$).

curs at a frequency of 1 kHz and at maximum rated power output. The *music power* rating of a hi-fi amplifier is defined as the peak power that can be delivered to the speaker for a very short period of time, with no more harmonic distortion than occurs at the maximum rated sine-wave rms power output level. That is, a music-power rating denotes the ability of an amplifier to process sudden peak musical wave forms without objectionable distortion. The peak duration that can be processed is limited by the capability of the filter capacitors in the power supply to sustain the required current demand.

## 3-2 Fundamental Amplifier Design Procedures

An example of a small-signal bipolar transistor audio-amplifier stage is diagrammed in Fig. 3-3. Small-signal operation denotes an input level from 1 microvolt ($\mu$V) to 10 millivolts (mV). A germanium 2N1414 PNP transistor is utilized. With the resistance values that are indicated, the transistor is forward biased for approximately 1 milliampere (mA) of collector; the transistor operates in class A and is connected in the common-emitter (CE) mode. The collector–emitter potential is approximately 6 V. A base–emitter input resistance of about 1,300 $\Omega$ is provided by the bias current. Note that this is not a dc resistance value; it is an ac resistance or dynamic resistance value. In other words, the applied ac signal $e_{in}$ "sees" a base–emitter resistance of about 1,300 $\Omega$. The dc resistance of the base–emitter junction is somewhat higher.

An emitter resistor is utilized in this configuration to provide temperature stabilization and thereby minimize changes in base-bias voltage

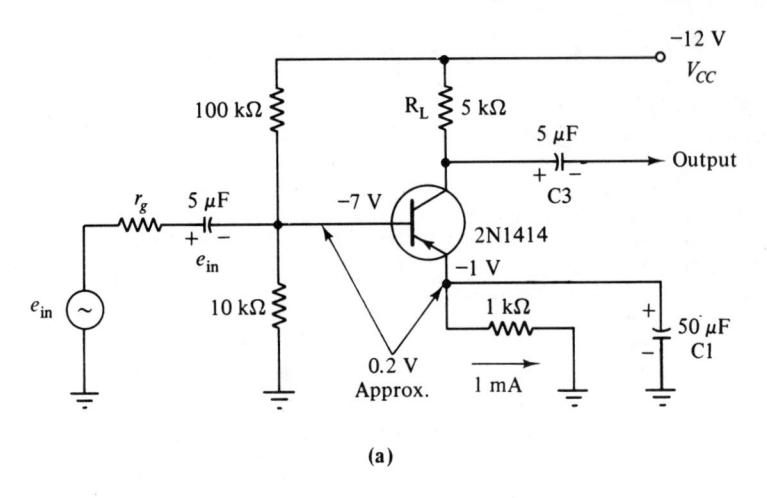

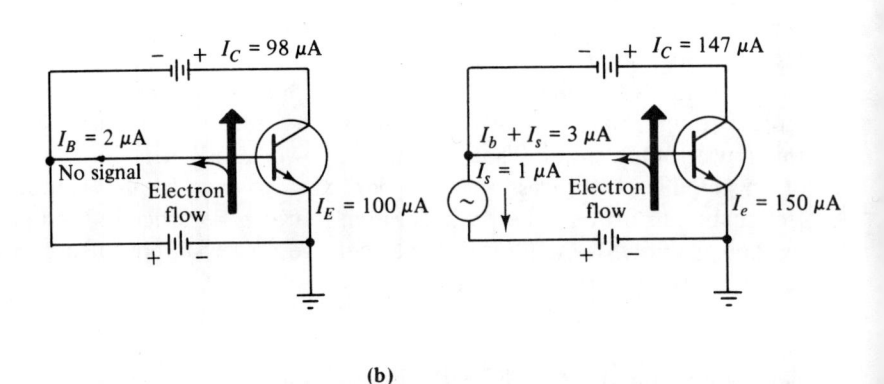

**Figure 3-3.** Example of a small-signal transistor audio-amplifier stage: (a) amplifier configuration; (b) basic transistor action.

versus temperature variation. That is, the emitter resistor provides dc current feedback; it does not provide ac negative feedback because it is bypassed by C1. Forward bias on the base is established by the 100-kΩ resistor connected from $V_{CC}$ to the base terminal. Because the value of this bias resistor is high, compared with the base–emitter resistance of the transistor, the bias circuit approximates a constant-current source. This design approach also assists in prevention of excessive current drain by the transistor as the ambient temperature increases. This is an example of an RC-coupled amplifier stage. That is, C2 couples the input signal into the base of the transistor, and C3 couples the collector output signal to the utilization circuit.

## 3-2 Fundamental Amplifier Design Procedures

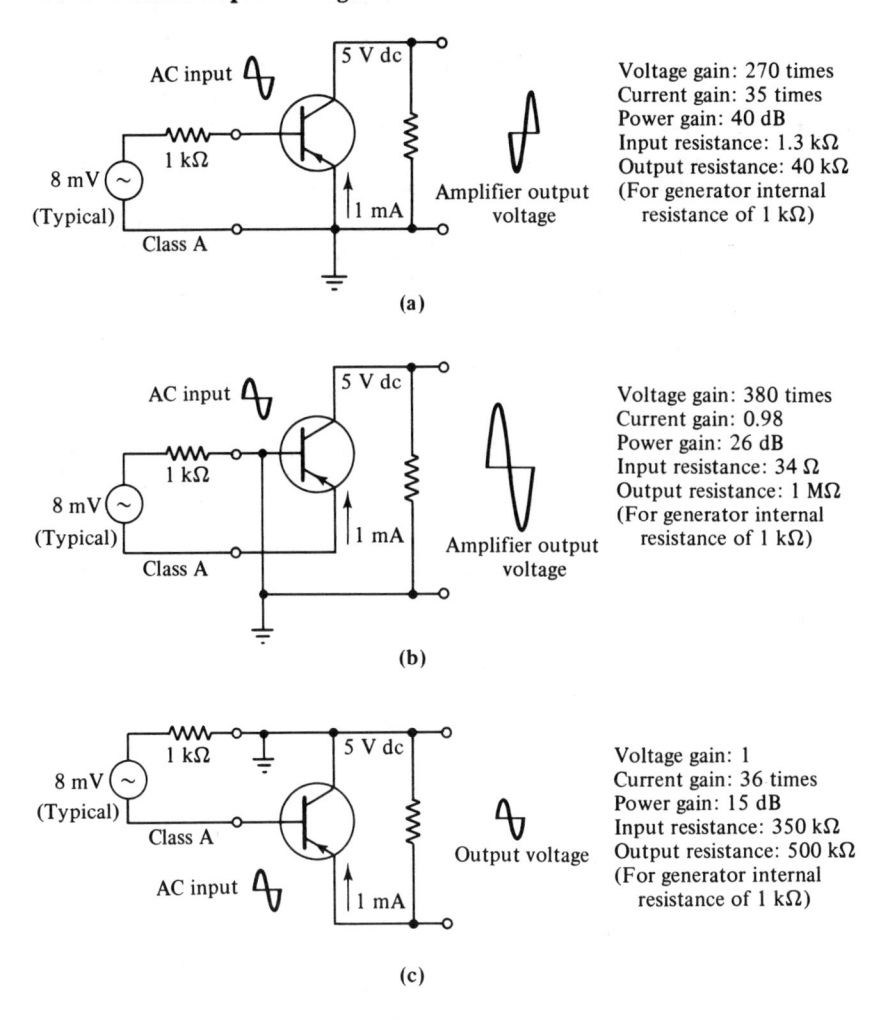

**(a)**

Voltage gain: 270 times
Current gain: 35 times
Power gain: 40 dB
Input resistance: 1.3 kΩ
Output resistance: 40 kΩ
(For generator internal resistance of 1 kΩ)

**(b)**

Voltage gain: 380 times
Current gain: 0.98
Power gain: 26 dB
Input resistance: 34 Ω
Output resistance: 1 MΩ
(For generator internal resistance of 1 kΩ)

**(c)**

Voltage gain: 1
Current gain: 36 times
Power gain: 15 dB
Input resistance: 350 kΩ
Output resistance: 500 kΩ
(For generator internal resistance of 1 kΩ)

Theoretical maximum efficiency = 50 per cent

**Figure 3-4.** Transistor circuit parameters for small-signal germanium transistors: (a) common emitter; (b) common base; (c) common collector.

Observe next the bipolar transistor circuit parameters listed in Fig. 3-4. The common-base (CB) configuration has the highest voltage gain; the common-collector (CC) configuration has the highest current gain; the common-emitter (CE) configuration has the highest power gain. The CB configuration has the lowest input resistance, and the CC configuration has the highest input resistance. The CB configuration has the highest output resistance, and the CC configuration has the lowest output resistance. Since the CE configuration has the least separated input- and output-resistance values, cascaded CE stages are more closely impedance matched than are cascaded CB or CC stages. Therefore, the CE configuration is widely employed by audio-circuit designers.

The theoretical maximum efficiency of a transistor operating in the class A mode is 50 percent. This fact follows from consideration of collector-circuit relations when the transistor is driven to its clipping and saturation points. In this condition of operation, the peak signal current is equal to the quiescent current value, and the signal power (for sine-wave operation) is equal to the dc power input. In turn, the operating efficiency of the transistor in the class A mode has a maximum value of 50 percent. In typical engineering practice, the operating efficiency is considerably less.

It is instructive to consider the effects of component and device tolerances on signal power output for the CE configuration. Refer to Fig. 3-5.

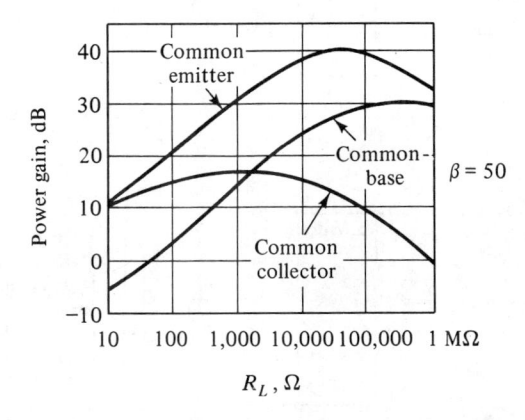

**Figure 3-5.** Power gain versus load resistance for a small-signal bipolar transistor.

It is seen that a ±20 percent tolerance on a 10,000-Ω load resistor will not result in a large tolerance on the signal-power output. However, the tolerance on the current-gain (beta) value of the transistor results in a large tolerance on the signal-power output. The power gain for an amplifier stage is equal to the product of its voltage gain and current gain. As an illustration, the power gain is proportional to the square of current, and is also proportional to the square of voltage. Accordingly, a doubling of the current gain, accompanied by a doubling of the voltage gain in a stage, will cause a quadrupling of the power gain. For example, consider a transistor with a beta value of 52.5 ± 17.7 percent. Its beta value may range from 35 to 70. If it is assumed that the stage operates in class A, a doubling in device current gain is accompanied by a doubling of stage current gain and by a doubling of stage voltage gain. Therefore, the power gain of the

## 3-2 Fundamental Amplifier Design Procedures

stage is quadrupled. In summary, a tolerance of ±17.5 percent on transistor beta corresponds to a tolerance of ±60 percent on the stage power output.

### Input Resistance Versus Load Resistance

As seen in Fig. 3-6, the input resistance of a CE stage is almost independent of the load-resistance value. However, the input resistance of a CB stage or of a CC stage will vary substantially from small to large values of load resistance. Observe that the horizontal axis in the diagram is basically numbered in $R_L/r_c$ units, where $R_L$ is the load-resistance value and $r_c$ is the "collector resistance" value of the transistor. This latter value is very large, on the order of 1 or more megohms. If we assume that the transistor has a beta value of 50, the horizontal axis may be alternatively numbered in $R_L$ units, as exemplified in Fig. 3-6. Note that the input resistance of various transistors is a function of frequency. As an illustration, a typical germanium PNP alloy drift-field transistor that has an input resistance of 1,350 Ω at low frequencies and up to 1.5 MHz exhibits an input resistance of only 150 Ω at 12.5 MHz.

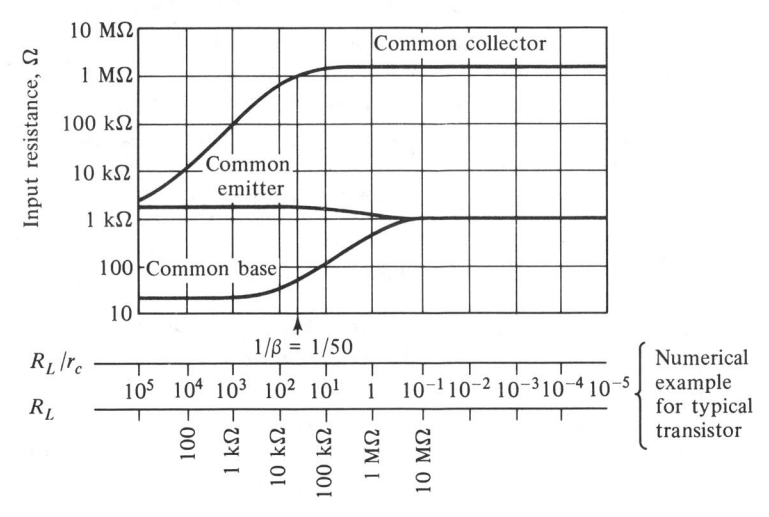

**Figure 3-6.** Input resistance versus load resistance for a small-signal bipolar transistor.

### Output Resistance Versus Generator Resistance

A CE stage has an output-resistance value that is reasonably independent of the generator-resistance value, as seen in Fig. 3-7. However, the output resistance of a CB stage, or of a CC stage, varies greatly from small to

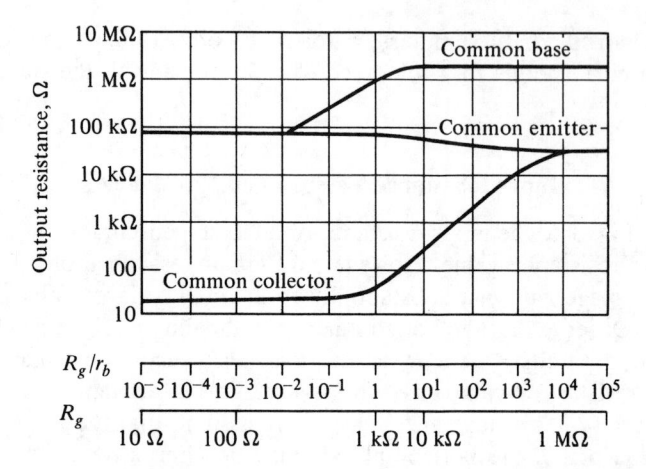

**Figure 3-7.** Output resistance versus generator resistance for a small-signal bipolar transistor.

large values of generator resistance. Observe that the horizontal axis in the diagram is basically numbered in $R_g/r_b$ units, where $R_g$ is the generator-resistance value and $r_b$ is the "base resistance" of the transistor. If we assume that the transistor has average characteristics, the horizontal axis may be alternatively numbered in $R_g$ units, as shown in Fig. 3-7. In addition to variation with generator resistance, the output resistance of various transistors is also a function of frequency. For example, a transistor that has an output resistance of 70 k$\Omega$ at low frequencies and up to 1.5 MHz exhibits an output resistance of only 4 k$\Omega$ at 12.5 MHz.

## Voltage and Current Amplification

When high values of load resistance are utilized, the voltage amplification of a stage operating in the CE, CB, or CC mode attains a plateau, as depicted in Fig. 3-8. On the other hand, at lower values of load resistance, the voltage amplification decreases. We observe that the CE and CB configurations have similar trends in voltage amplification versus load resistance, although the gain of a CE stage is slightly higher than that of a CB stage for low values of load resistance. As would be anticipated, the voltage amplification of a CC stage is practically unity for all except very low values of load resistance. At very low load resistance values, the transistor works into an approximate short-circuit load condition, and the voltage amplification decreases. It may be observed that uniformity of voltage amplification does not correspond to uniformity of power amplification in any of three operating modes. That is, the current amplification

## 3-2 Fundamental Amplifier Design Procedures

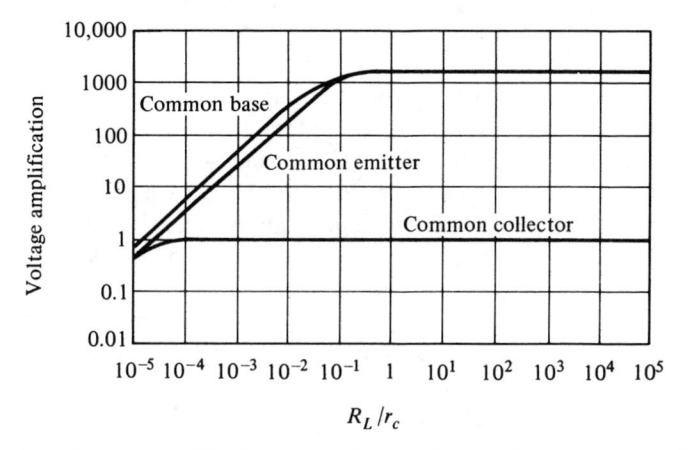

**Figure 3-8.** Voltage amplification versus load resistance for a small-signal bipolar transistor.

of a stage varies most extensively over the load-resistance range where the voltage amplification varies the least, as explained below.

With reference to Fig. 3-9, the current-amplification characteristic versus load resistance of a small-signal bipolar transistor is shown. In comparison with the voltage-amplification characteristic, current amplification of a stage attains a plateau at low values of load resistance. Because the power amplification of a stage is equal to the product of its voltage amplification and its current amplification, the power output attains its maximum value when a suitable value of load resistance is employed, as was shown in Fig. 3-5. Different values of load resistance are required to obtain

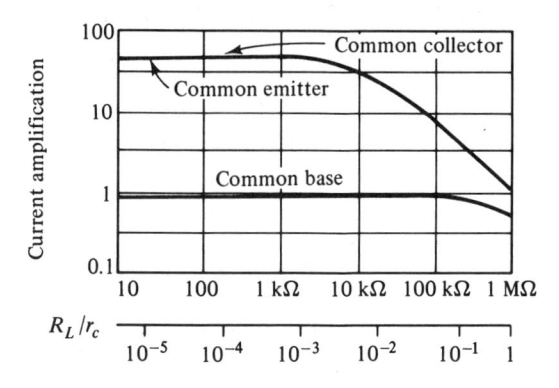

**Figure 3-9.** Current amplification versus load resistance for a small-signal bipolar transistor.

maximum power amplification in the CE, CB, and CC modes of operation. In many audio-circuit design projects, the chief requirement in stage operation is to realize maximum power output.

Both the voltage amplification and the current amplification of a stage will vary in accordance with the tolerance on the beta value of the transistor. However, tolerances on the load-resistance value have much less effect on current and voltage amplification. Transistors in most production lots have a comparatively wide tolerance on beta value. In turn, the audio-circuit designer ordinarily chooses somewhat elaborated amplifier circuitry that tends to minimize the lack of uniformity in beta values (see Fig. 3-10). In some applications, such as in complementary-symmetry output amplifiers, pairs of transistors must have beta values that are practically the same if the stage is to perform satisfactorily. Such selected transistors are called *matched pairs*.

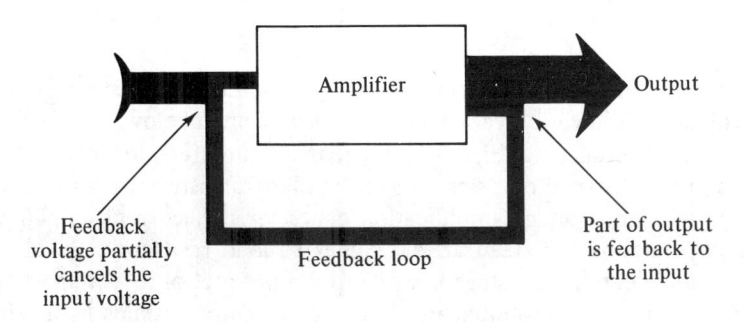

**Figure 3-10.** Basic principle of negative feedback.

## 3-3 Preamplification Versus Power Amplification

Most high-fidelity amplifiers can be grouped into preamplifier or power-amplifier categories. A preamplifier may be designed as a separate unit, or it may be designed as an integral part of a complete audio amplifier. A 12-W high-fidelity power amplifier is illustrated in Fig. 3-11. Its efficiency at maximum power output is 67 percent. A power amplifier of this type requires approximately 1 V of drive signal from a preamplifier. A typical preamplifier arrangement is depicted in Fig. 3-12. It consists of three stages that operate in class A. The maximum available voltage gain is 69 dB, with substantially less than 1 percent total harmonic distortion (THD). Four input ports are provided to accommodate a low-impedance microphone, a tape player, a record player, and an FM tuner. Highest sensitivity is provided for the low-Z microphone input; in other words, an input voltage of 350 $\mu$V will develop 1-V output into a load of 10 k$\Omega$.

## 3-3 Preamplification Versus Power Amplification

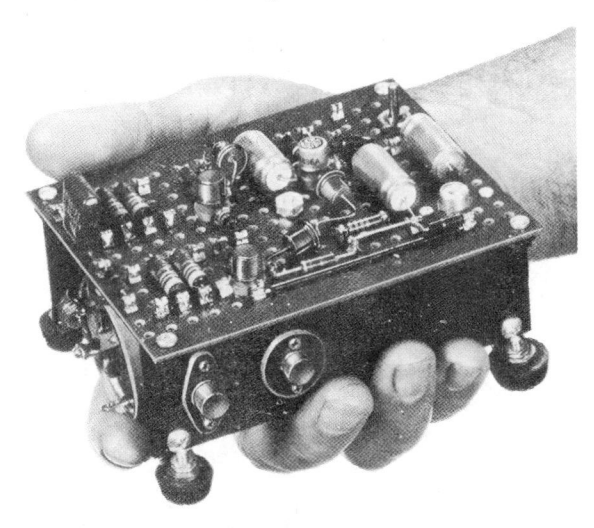

**Figure 3-11.** A 12-W high-fidelity amplifier. (*Courtesy, GE*)

The tape input is designed for lower sensitivity; an input of 1.5 mV will drive the preamplifier to 1-V output. Similarly, the phono input requires a still higher input level of 8 mV; the FM tuner input operates normally at an input level of 250 mV. A small-signal transistor is rated for a maximum power dissipation of less than 1 W. On the other hand, a power transistor is rated for a maximum power dissipation in the range from 1 to 40 W.

Next consider the typical voltage-gain and power-output values for a power amplifier, as noted in Fig. 3-13. A signal-voltage gain of 18 dB is provided; an input of 1 V rms develops an output level of approximately 8 V rms across an 8- or 16-Ω load. Since the input impedance of the power amplifier is much higher than its output or load impedance (10 kΩ versus 8 Ω), its power gain is considerable although its voltage gain is only moderate. On the other hand, a preamplifier is characterized by high voltage gain and only moderate power gain. A power amplifier is rated for frequency response within ±1 dB at maximum power output, and for frequency response within ±3 dB at maximum power output. Many power amplifiers are also rated for high-fidelity power bandwidth, as depicted in Fig. 3-14.

Power bandwidth is characterized as follows. A power amplifier has a maximum rated power output, such as 12 W. Note that if a power amplifier is operated at a level in excess of its maximum rated output, its output transistors will overheat and become damaged. When tested at a frequency of 1 kHz, the amplifier will develop a certain amount of distortion. For example, the THD at 1 kHz and maximum power output

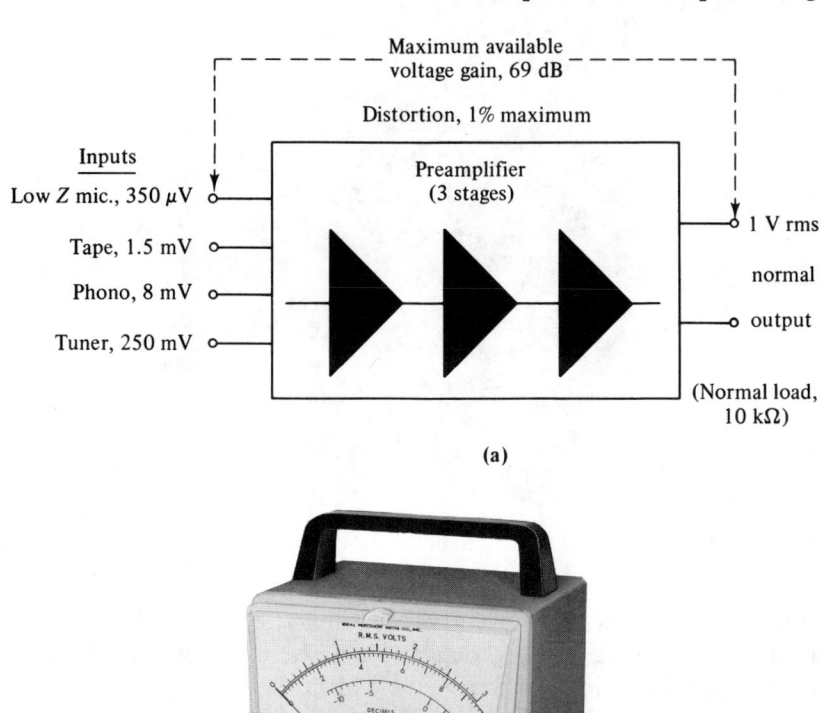

**Figure 3-12.** Typical voltage gain values for a hi-fi preamplifier: (a) input–output levels; (b) an audio voltmeter for measurement of low-level signal voltages. (*Courtesy, Heath Co.*)

might be 1 percent. Now, if the drive to the amplifier is reduced so that the output level is decreased by 3 dB, the percentage distortion will decrease. With the drive level held constant, and as the test frequency reduces toward a lower limit, a point will be reached at which the percentage distortion rises to 1 percent. This is the lower limit of the power bandwidth.

## 3-3 Preamplification Versus Power Amplification

Similarly, with the drive level held constant, and as the test frequency increases toward an upper limit, a point will be reached at which the percentage distortion rises to 1 percent. This is the upper limit of the power bandwidth. Typical limiting frequencies are indicated in Fig. 3-14.

The theoretical maximum power efficiency of a class B amplifier is

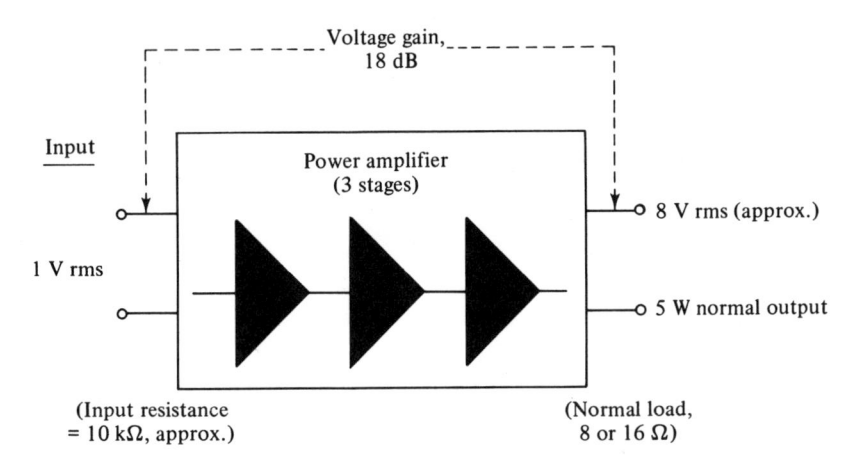

**Figure 3-13.** Typical voltage-gain and power-output values for a hi-fi audio power amplifier.

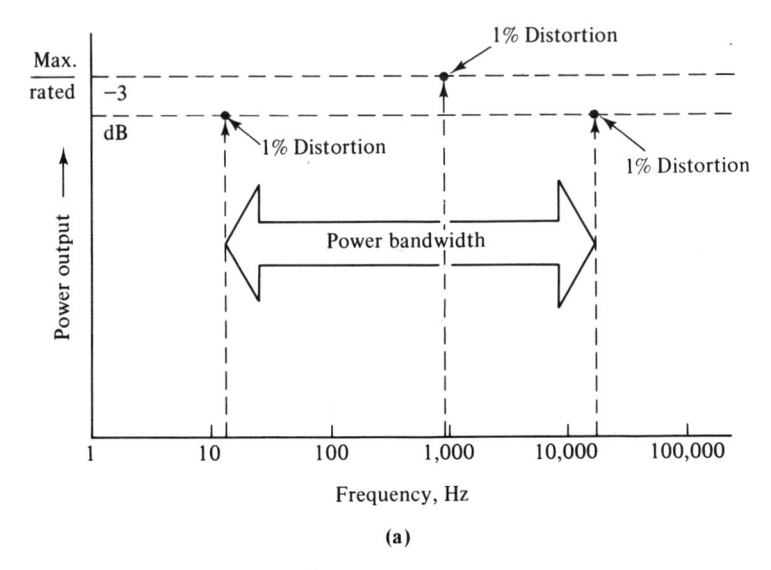

**Figure 3-14.** High-fidelity power bandwidth: (a) definition.

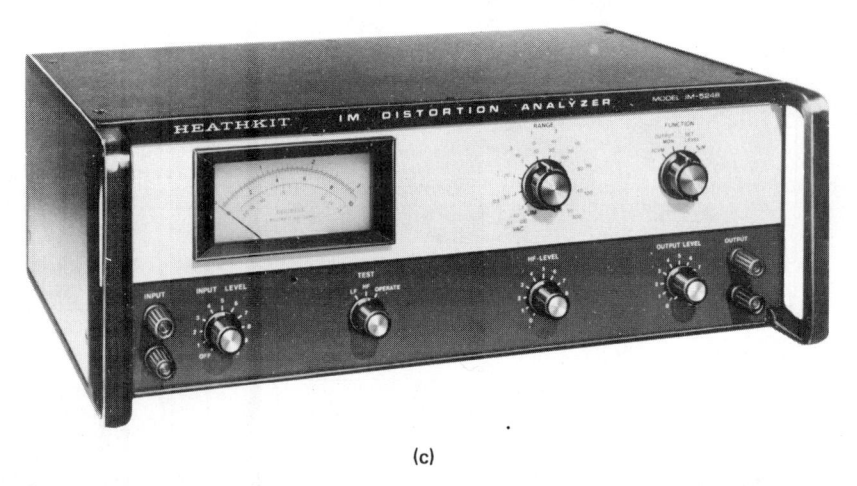

**Figure 3-14.** (*Continued*) (b) harmonic distortion meter; (c) intermodulation distortion analyzer. (*Courtesy, Heath Co.*)

78.5 percent. In actual practice, the efficiency is appreciably less; an efficiency of 60 percent is typical. If the amplifier is operated in class AB, its efficiency becomes intermediate to that of a class A or class B amplifier.

## 3-4 Transistor Load Lines

Transistors are inherently nonlinear devices, as seen in Fig. 3-15. In other words, the base-current/base-voltage characteristics are not linear; they become progressively curved at smaller base-current values. Moreover,

## 3-4 Transistor Load Lines

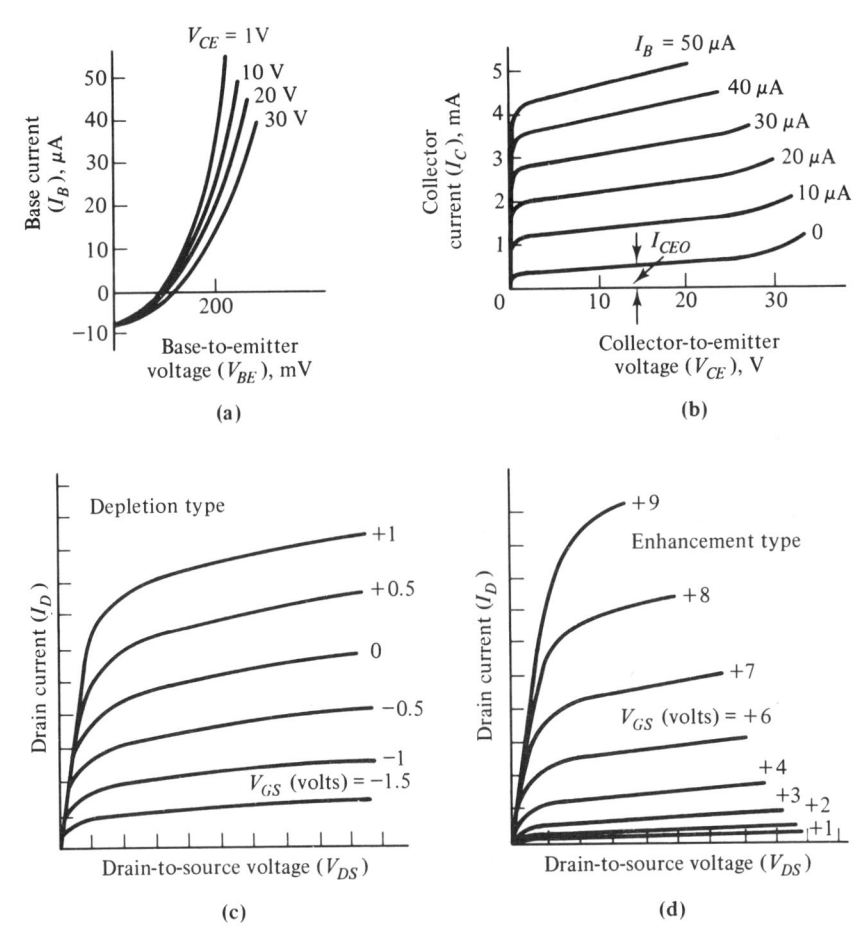

**Figure 3-15.** Transistor characteristics: (a) base family characteristics for bipolar transistor in CE mode; (b) collector family characteristics for bipolar transistor in CE mode; (c) drain family characteristics for unipolar transistor (MOSFET) in common-source mode (depletion type); (d) drain family characteristics for unipolar transistor (MOSFET) in common-source mode (enhancement type).

the collector characteristics become more or less "cramped" at higher collector-current values. These nonlinear relations between voltage and current values are the basis of harmonic distortion in amplifier operation. It is a basic principle of nonlinear circuit operation that distortion decreases as the signal level decreases. In turn, the audio-circuit designer is sometimes justified in treating a transistor as a linear device, provided that a low signal level is maintained (small-signal operation). This is typically the case for the input stage of a preamplifier. On the other hand, a power amplifier operates at a high signal level, and it cannot be assumed that a power transistor is a linear device. Calculations are comparatively difficult in nonlinear circuitry; however, graphical constructions facilitate circuit design procedures.

Consider the simple circuit comprising both linear and nonlinear resistance depicted in Fig. 3-16. We will solve the circuit for the following data:

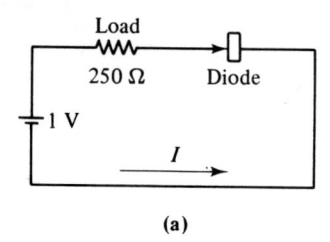

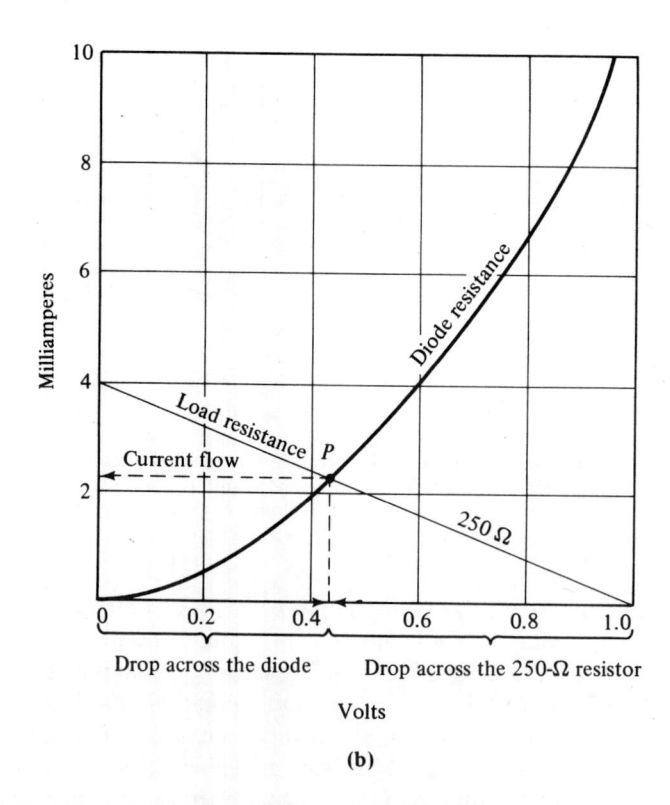

**Figure 3-16.** Graphical solution of current value and voltage drops in a diode circuit: (a) circuit. (b) placement of load line.

## 3-4 Transistor Load Lines

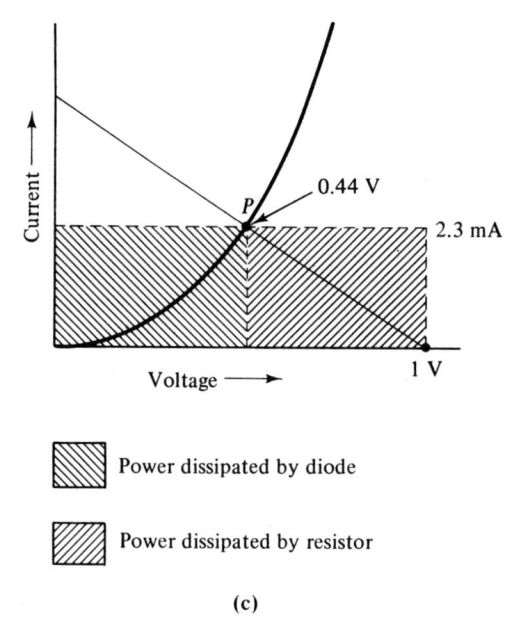

[Figure: graph showing current vs voltage with diode characteristic curve and load line intersecting at point P = 0.44 V, 2.3 mA; load line endpoint at 1 V. Legend shows shaded regions for "Power dissipated by diode" and "Power dissipated by resistor".]

(c)

**Figure 3-16.** (*Continued*) (c) determination of power values.

1. What is the current flow in the circuit?
2. What is the voltage drop across the diode?
3. What is the voltage drop across the resistor?
4. How much power does the diode dissipate?
5. How much power does the resistor dissipate?
6. How much power does the battery supply to the circuit?

A voltage–current characteristic of the diode is utilized for the graphical construction. A load line is drawn, as shown in Fig. 3-16a. This load line defines the resistance value of the linear resistor in terms of its voltage drop and current flow. That is, if the resistor drops 1 V, the current flow through the circuit must be 4 mA; or, if the resistor drops 0 V, the current flow will be zero. These limiting values are indicated by the ends of the load line. In effect, the load line and the diode characteristic represent a pair of simultaneous equations with respect to Ohm's law. Accordingly, the operating point (quiescent point) for the circuit occurs at the intersection of the load line with the diode characteristic. In this example, the quiescent point (*P*) corresponds to a current flow of 2.3 mA, approximately. The voltage drop across the diode is 0.44 V, approximately, and the voltage drop across the linear resistor is 0.56 V, approximately. These two drops add to 1 V, the source-voltage value, in accordance with Kirchhoff's law.

With reference to Fig. 3-16c, the areas determined by point $P$, the 0-V point, and the 1-V point on the abscissa define the power relations in the circuit. In other words, the area enclosed by the voltage and current axes up to point $P$ represents the power dissipated by the diode. Similarly, the remaining area up to the 1-V limit represents the power dissipated by the resistor in the circuit. Otherwise stated, the area of a rectangle is equal to the length of its base multiplied by the length of its altitude. Or these areas are proportional to the products of voltage drops and current flow. Thus the power dissipated by the diode is equal to 1.01 mW, approximately, and the power dissipated by the resistor is equal to 1.29 mW, approximately. In accordance with Joule's law, the power supplied by the battery is equal to the sum of these two power-dissipation values, or 2.3 mW.

This same principle is utilized in determining the operation of a transistor amplifier stage, as exemplified in Fig 3-17. A load line is drawn on the collector family of transistor characteristics. In this example, the collector load resistor has a value of 3,000 Ω. Therefore, the load line joins the 12-V and 4-mA points on the diagram. Adjustment of the base–emitter bias voltage to some point approximately midway between cutoff and saturation will place the transistor in its *active,* or *linear,* region of operation. When operating within this region, the transistor functions as an amplifier. If the bias voltage is adjusted for a quiescent point at $B$ (2 mA), class A amplification is obtained. This corresponds to a base current of 65 μA, approximately.

Observe that if an input signal is applied to the amplifier arrangement depicted in Fig. 3-17 the base current varies. In this example, the swing along the load line is between points $A$ and $C$; the base current varies from approximately 10 to 125 μA. In turn, the collector current swings from 0.5 to 3.3 mA, approximately. This variation in collector current produces a voltage swing across the load resistor of 8.5 V, approximately. Thus the stage has a current gain of about 24 times. The power output from the stage for the exemplified base drive is 24 mW, approximately. Note in passing that this transistor has an average beta value of about 28. In the quiescent condition (operating point at $B$), the transistor dissipates approximately 10 mW, and the load resistor dissipates about 14 mW.

If the audio circuit designer desires to obtain a comparatively high input impedance to a stage, he may choose a junction-field-effect transistor (JFET) or a metal-oxide semiconductor field-effect transistor (MOSFET). Comparative device parameters are listed in Table 3-2. It is desirable to utilize a low-noise type of transistor in an input stage, because the input stage contributes more noise to the output signal than does any following

## 3-4 Transistor Load Lines

**Common-emitter mode**

Saturation region (transistor turned on: $I_B$ high, $I_C$ high, $V_{CE}$ low)

Collector current $I_C$, mA

$I_B = 150\ \mu A$

$I_B = 100\ \mu A$

$I_B = 50\ \mu A$

$I_B = 25\ \mu A$

$I_B = 0$

Active (transition) region (transistor amplifies: $I_B$, $I_C$, and $V_{CE}$ swing with signal input between points A and C. B is no signal, dc bias point)

Collector voltage $V_{CE}$, V

Cutoff region (transistor turned off: $I_B$ low, $I_C$ low, $V_{CE}$ high)

(a)

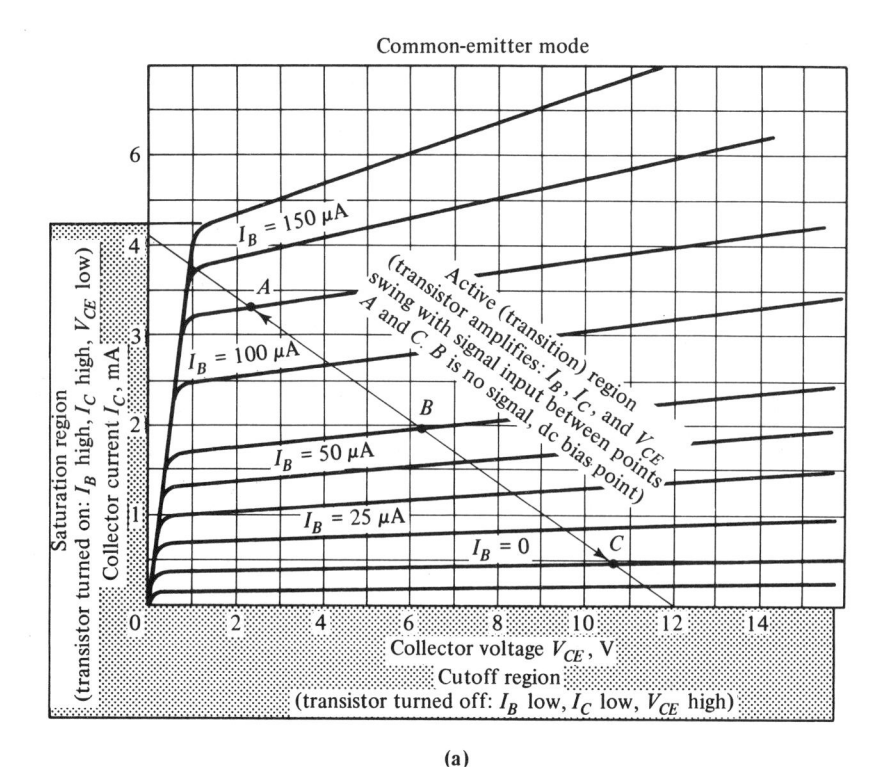

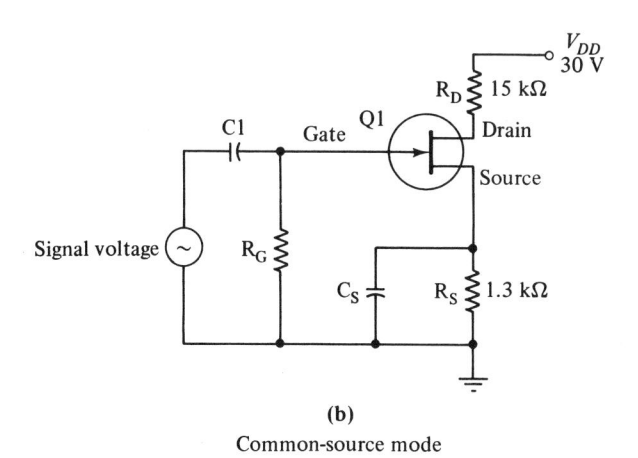

(b)
Common-source mode

$V_{DD}$ 30 V
$R_D$ 15 kΩ
C1 Gate Q1 Drain
Source
Signal voltage $R_G$
$C_S$ $R_S$ 1.3 kΩ

**Figure 3-17.** Typical transistor characteristics, with load lines: (a) bipolar collector family with a 3,000-Ω load line; (b) JFET common-source amplifier configuration.

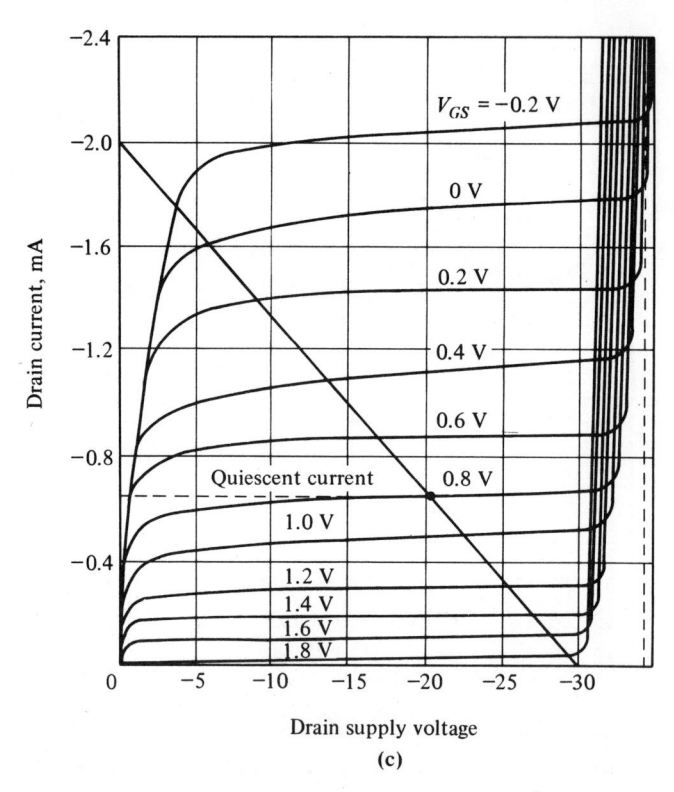

**Figure 3-17.** (*Continued*) (c) JFET drain family, with 15-kΩ load line.

Table 3-2.   Comparative Device Parameters

| Parameter | Bipolar Transistor | JFET Transistor | MOSFET Transistor |
|---|---|---|---|
| Input impedance | Low | High | Very high |
| Noise | Low | Low | Unpredictable |
| Aging | Not noticeable | Not noticeable | Noticeable |
| Bias voltage temperature coefficient | Low and predictable | Low and predictable | High and unpredictable |
| Control electrode current | High | 0.1 nA | 10 pA |
| Overload capability | Comparatively good | Comparatively good | Poor |

## 3-4 Transistor Load Lines

**Table 3-2.** (*Continued*)

Input Impedance (Resistance) of Common-Collector Stage

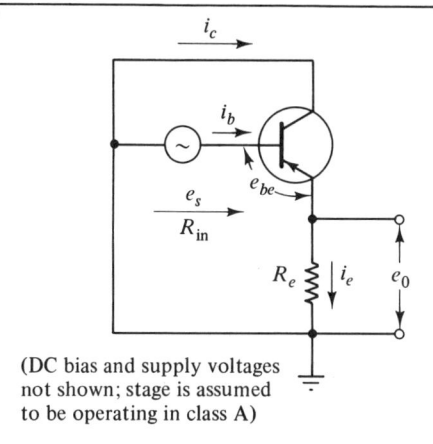

(DC bias and supply voltages not shown; stage is assumed to be operating in class A)

The ac input impedance, or resistance $R_{in}$, is very high in the common-collector configuration because of the negative feedback action in the emitter circuit. Consider that $R_e = 0$; then $e_0 = 0$, $e_{be} = e_s$, and $i_b$ has its maximum value. Next, if $R_e$ has a high resistance value, $e_0$ will almost as large as $e_s$, and $e_{be}$ will be small; $e_{be}$ is equal to the difference between $e_s$ and $e_0$. Since $e_{be}$ is small, $R_{in}$ becomes large; $R_{in}$ is equal to $e_s/i_b$, in accordance with Ohm's law.

Input Impedance (Resistance) of Common-Emitter Stage with Emitter Current Feedback

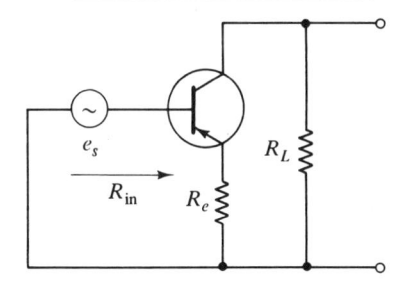

(DC bias and supply voltages not shown; stage is assumed to be operating in class A)

The ac input impedance, or resistance $R_{in}$, is rather high in the common-emitter configuration when there is substantial negative-feedback action in the emitter circuit. A common-emitter configuration with emitter feedback has an input impedance (resistance) that is higher than that of a CE stage without emitter feedback, but lower than that of the common-collector configuration. This relation follows from the condition of less than 100 percent negative feedback in the CE configuration with an unbypassed emitter resistor.

## Table 3-2. (*Continued*)

Output Impedance (Resistance) of Common-Emitter Stage
with Emitter Current Feedback

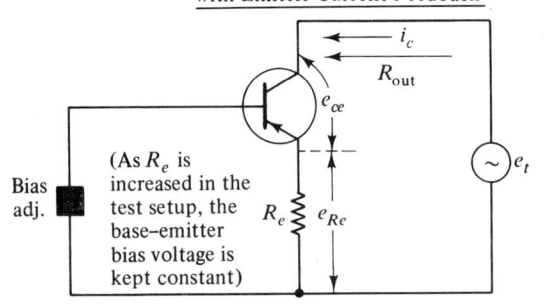

(As $R_e$ is increased in the test setup, the base–emitter bias voltage is kept constant)

(DC bias and supply voltages not shown; stage is assumed to be operating in class A)

The ac output impedance, or resistance $R_{out}$, is rather high in the common-emitter configuration when there is substantial negative-feedback action in the emitter circuit. Consider that $R_e = 0$; then $e_{ce}$ will be equal to the test voltage $e_t$. Next, if $R_e$ has a high resistance value, $e_{Re}$ will be high; in turn, $e_{ce}$ becomes smaller, because $e_{ce}$ is equal to the difference between $e_t$ and $e_{Re}$. Consequently, $i_c$ becomes smaller and $R_{out}$ becomes higher, because $R_{out}$ is equal to $e_t/i_c$ in accordance with Ohm's law.

Input Impedance (Resistance) of Common-Emitter Stage
with Voltage Feedback

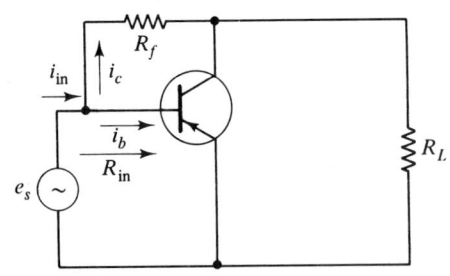

(DC bias and supply voltages not shown; stage is assumed to be operating in class A)

The ac input impedance, or resistance $R_{in}$, is rather low in the common-emitter configuration when voltage feedback is used. Consider that $R_f = \infty$; then the input resistance is equal to $e_s/i_b$. Next, if $R_f$ has a moderate value, the total input current $i_{in}$ becomes greater; $i_{in}$ is then equal to $i_b + i_c$, where $i_c$ is the "cancellation current" due to mixture of $e_s$ with the out-of-phase feedback voltage. Since $i_{in}$ is greater when negative feedback is used, the input resistance $R_{in}$ decreases, because $R_{in} = e_s/i_{in}$, according to Ohm's law.

## Table 3-2. (*Continued*)

Output Impedance (Resistance) of Common-Emitter Stage with Voltage Feedback

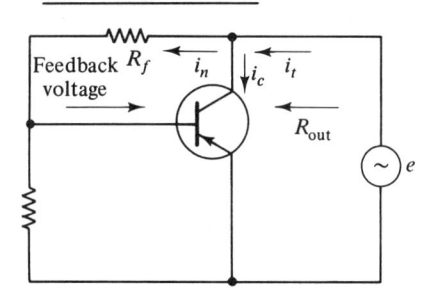

(DC bias and supply voltages not shown; stage is assumed to be operating in class A)

The ac output impedance or resistance $R_{out}$, is rather low in the common-emitter configuration when voltage feedback is used. Consider that $R_f = \infty$; then the output resistance is equal to $e/i_c$. Next, if $R_f$ has a moderate value, the total current $i_t$ becomes greater; $i_t$ is then equal to $i_c + i_n$, where $i_n$ is the "neutralization" or cancellation current due to mixture of $e$ with the amplified feedback voltage that appears at the collector. Since $i_t$ is greater when negative feedback is used, the output resistance $R_{out}$ decreases, because $R_{out} = e/i_t$, in accordance with Ohm's law.

stage. Overload capability is often a basic consideration also, inasmuch as the designer has no control of the conditions of use. Although a simple MOSFET transistor has poor overload capability, the circuit designer may specify a gate-protected type of MOSFET (Fig. 3-18); this built-in protective feature ensures that the transistor will withstand any reasonable overload to which it may be subjected.

Transistor amplification is essentially linear, provided that the device is operated in class A and that the signal swing is quite small. On the other

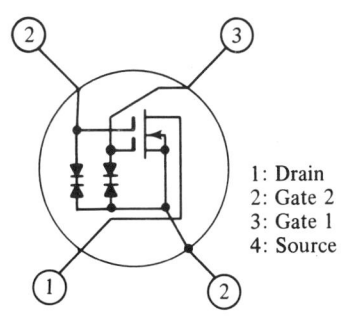

1: Drain
2: Gate 2
3: Gate 1
4: Source

**Figure 3-18.** Dual gate-protected MOSFET.

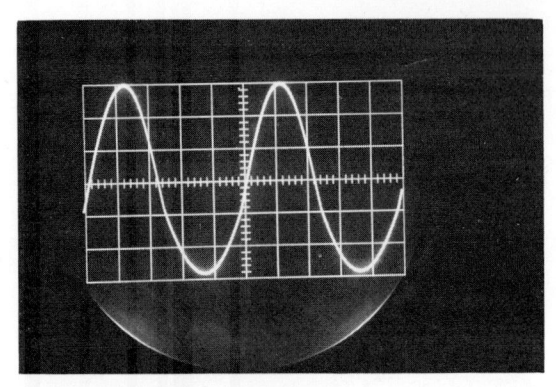

**Figure 3-19.** Example of nonlinear amplification.

hand, a bias shift from class A operation and/or application of a comparatively large signal swing results in a distorted output wave form, as exemplified in Fig. 3-19. Observe that fewer horizontal baseline intervals are included by the positive excursion of the waveform than by its negative excursion. This is an example of harmonic distortion; it can be minimized by negative-feedback action, as explained next.

## 3-5 Negative-Feedback Principles

Inherent nonlinearity is evident in the transfer function of a transistor, as exemplified in Fig. 3-20. This is a graph of collector current versus base–emitter voltage for a small-signal germanium transistor. If the transistor operates at a bias of 0.2 V, it is apparent that the input signal swing must be very limited in order to approximate a linear transfer function. If a

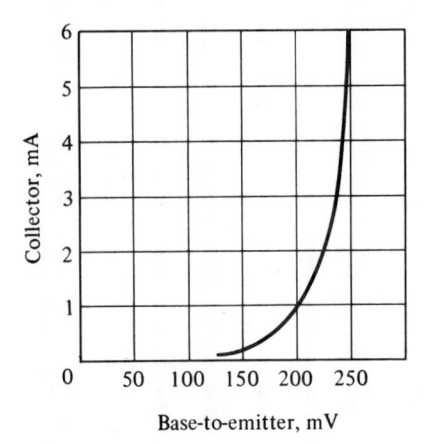

**Figure 3-20.** Transfer characteristic for a small-signal transistor.

## 3-5 Negative-Feedback Principles

comparatively large input signal swing were employed, such as from 175 to 225 mV, the output wave form from the collector would be seriously distorted, unless negative feedback were utilized. Predistortion is the basis of negative-feedback action, as seen in Fig. 3-21. In other words, an amplifier circuit arrangement is used that predistorts the signal into the base of the transistor in a manner such that the output signal has much less distortion than otherwise. Thus negative feedback does not linearize the action of the transistor; it functions to compensate for transistor nonlinearity.

A portion of the output signal is fed back and mixed out-of-phase with the input signal, so that partial cancellation of the source signal voltage occurs. In this process, a difference signal is produced that is a predistorted version of the source signal. The predistorted signal is mixed with a greater or less percentage of undistorted signal voltage, and is ap-

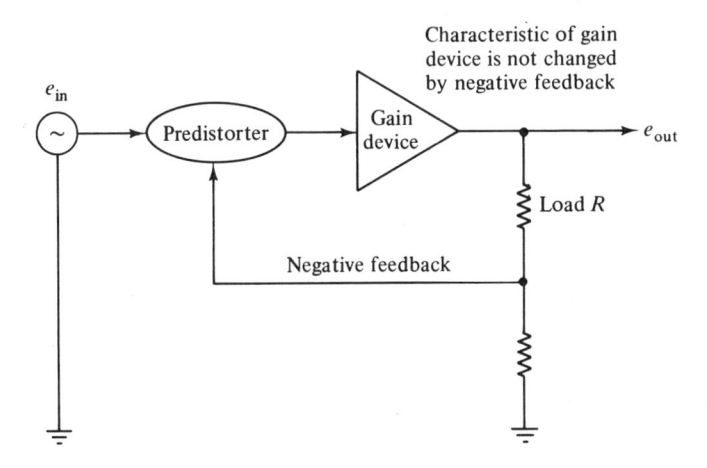

**Figure 3-21.** Predistortion is the basis of negative-feedback action.

plied to the base of the transistor. If a sufficiently large amount of signal is fed back through the negative-feedback circuit, the transistor can, in effect, be made as nearly distortionless as desired. Of course, there is a tradeoff involved in negative-feedback operation. If the audio circuit designer employs a very large amount of negative feedback, he must increase the source-signal-voltage level accordingly, in order to obtain the same output signal level as in the absence of negative feedback. Or, if the source-signal-voltage level cannot be increased, the circuit designer must include additional amplifier stages to obtain the same output signal level.

With reference to Fig. 3-22, consider the voltage levels at the key points in the negative-feedback amplifier arrangement. Since a 10-mV input signal provides a 2-V output signal, the intrinsic gain of the amplifier is 200 times. Next note that the output voltage drives the negative-feedback circuit. In this example, the output from the feedback network is 90 mV. Feedback takes place out of phase with the input signal voltage, so that 10 mV is subtracted from the 100-mV source, leaving 10 mV to drive the amplifier transistor. Or a source signal voltage of 100 mV produces an output signal voltage of 2 V. In effect, the stage gain is 20 times instead of 200 times because of negative-feedback action. Although this is a large reduction from the intrinsic gain value, it can represent an acceptable tradeoff for the audio circuit designer because of its incidental improvement in fidelity.

Observe in Fig. 3-22 that the source signal voltage is reduced to 10 percent of its original value at the input to the transistor. In turn, a circuit designer says that 20 dB of negative feedback is employed in this example. An amplifier that utilizes 20 dB or more of negative feedback is said to have *significant* feedback. This term implies that the feedback action is sufficient to reduce distortion substantially; for example, if the transistor has an inherent distortion of 5 percent without negative feedback, 20 dB of negative feedback will reduce this distortion to 0.5 percent, approximately. Note in passing, however, that there is one type of distortion that cannot be reduced by the use of negative feedback. This exception occurs in the case of clipping distortion, as depicted in Fig. 3-23. Negative-feedback action fails to improve the fidelity of the stage in this situation, because the transistor has no reserve gain past its collector-current cutoff point.

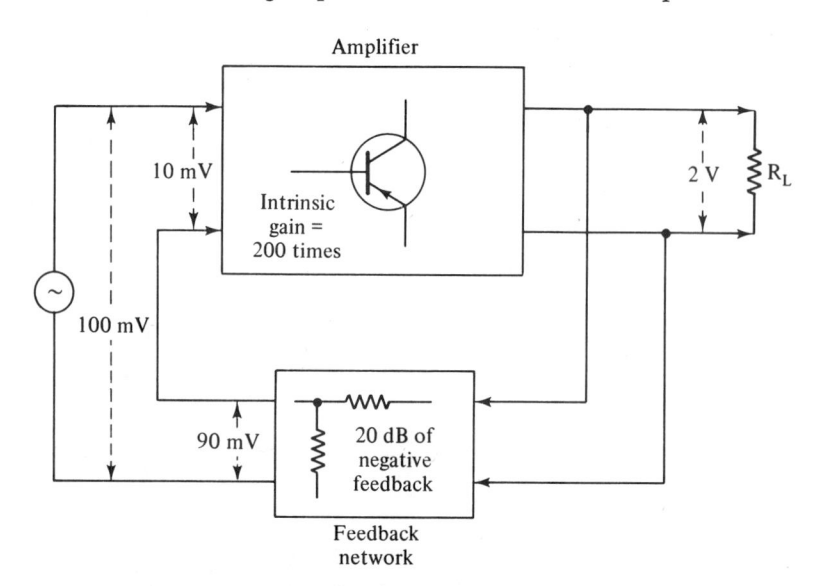

**Figure 3-22.** Example of negative-feedback circuit relations.

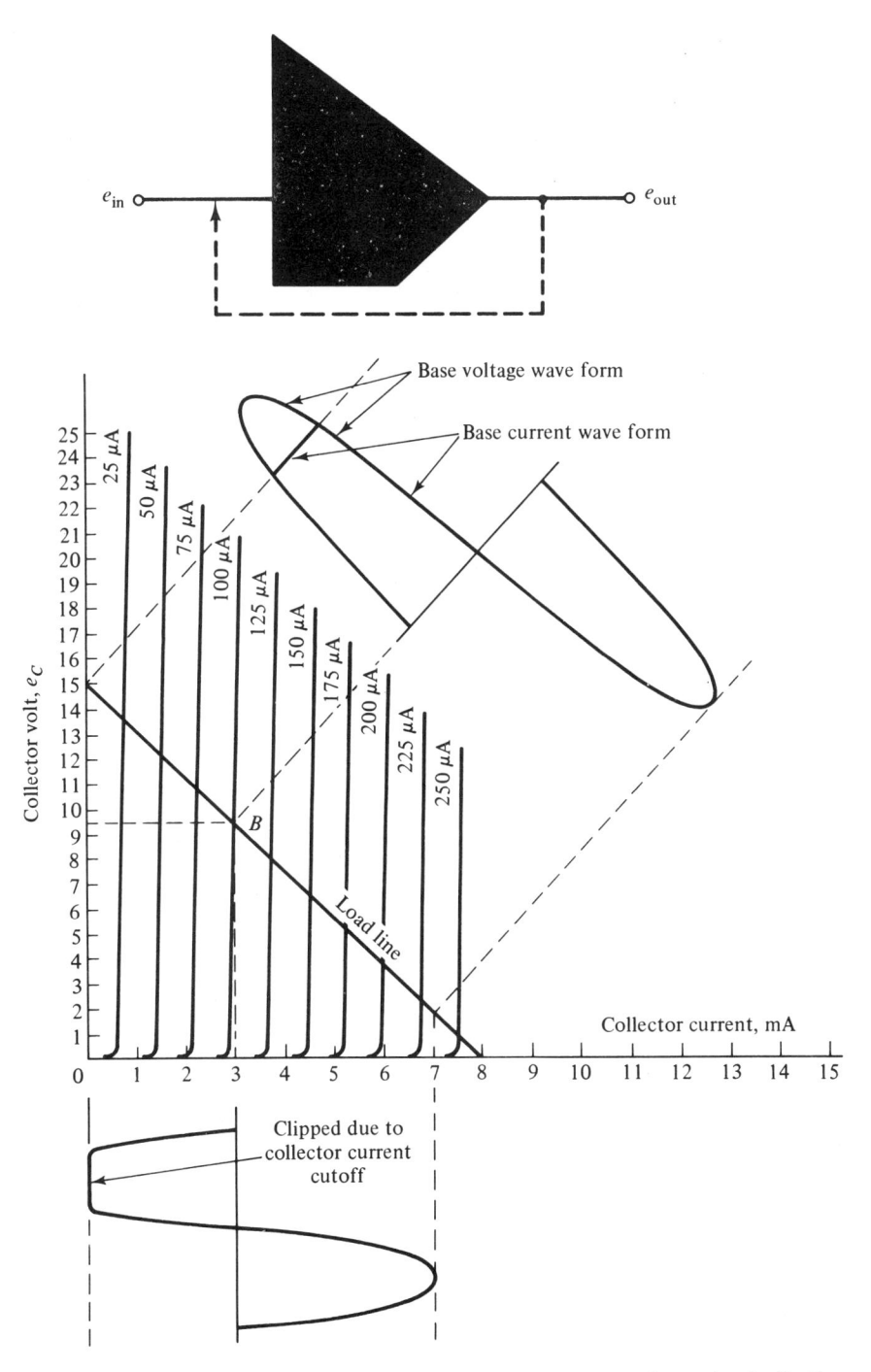

**Figure 3-23.** Clipping distortion cannot be corrected by means of negative feedback.

An important benefit of negative-feedback action is that tolerances on transistors have less influence on amplifier action than in a similar configuration without negative feedback. As an illustration, suppose that the transistor in the arrangement of Fig. 3-22 is replaced by a transistor that has only 50 percent of the gain of the first transistor. In the absence of negative feedback, only half the original output voltage would be obtained. On the other hand, when 20 dB of negative feedback is utilized, the output voltage drops only 10 percent. That is, 90 percent of the original output amplitude is obtained, although the replacement transistor has only one half the gain of the original transistor. In effect, 20 dB of negative feedback has reduced a 50 percent tolerance on transistor gain to an effective tolerance of 10 percent. This improvement in effective tolerance can ensure that a production run of amplifiers will have a reasonably restricted range of amplification variation, despite comparatively wide tolerance on the transistor gain value.

As shown in Fig. 2-15, four prototype negative-feedback arrangements are used by audio circuit designers:

1. *Series–parallel configuration.* The amplifier input and the feedback network are connected in series, whereas the amplifier output and the feedback network are connected in parallel.
2. *Parallel–parallel configuration.* The amplifier input and the feedback network are connected in parallel, and the amplifier output and the feedback network are also connected in parallel.
3. *Series–series configuration.* The amplifier input and the feedback network are connected in series, and the amplifier output and the feedback network are also connected in series.
4. *Parallel–series configuration.* The amplifier input and the feedback network are connected in parallel, whereas the amplifier output and the feedback network are connected in series.

*Voltage feedback* is employed in the configurations of Fig. 3-24a and Fig. 3-24b; in this arrangement, the output voltage tends to remain independent of the value of $R_L$. *Current feedback* is utilized in the configurations of Fig. 3-24c and Fig. 3-24d; in this arrangement, the output current tends to remain independent of the value of $R_L$. That is, a parallel-output connection tends to make an amplifier operate as a constant-voltage source, whereas a series-output connection tends to make an amplifier operate as a constant-current source. Otherwise stated, a parallel-output connection decreases the amplifier's output resistance. On the other hand, a series-output connection increases the amplifier's output resistance.

From another useful viewpoint, two basic feedback arrangements

## 3-5 Negative-Feedback Principles

**(a)**

**(b)**

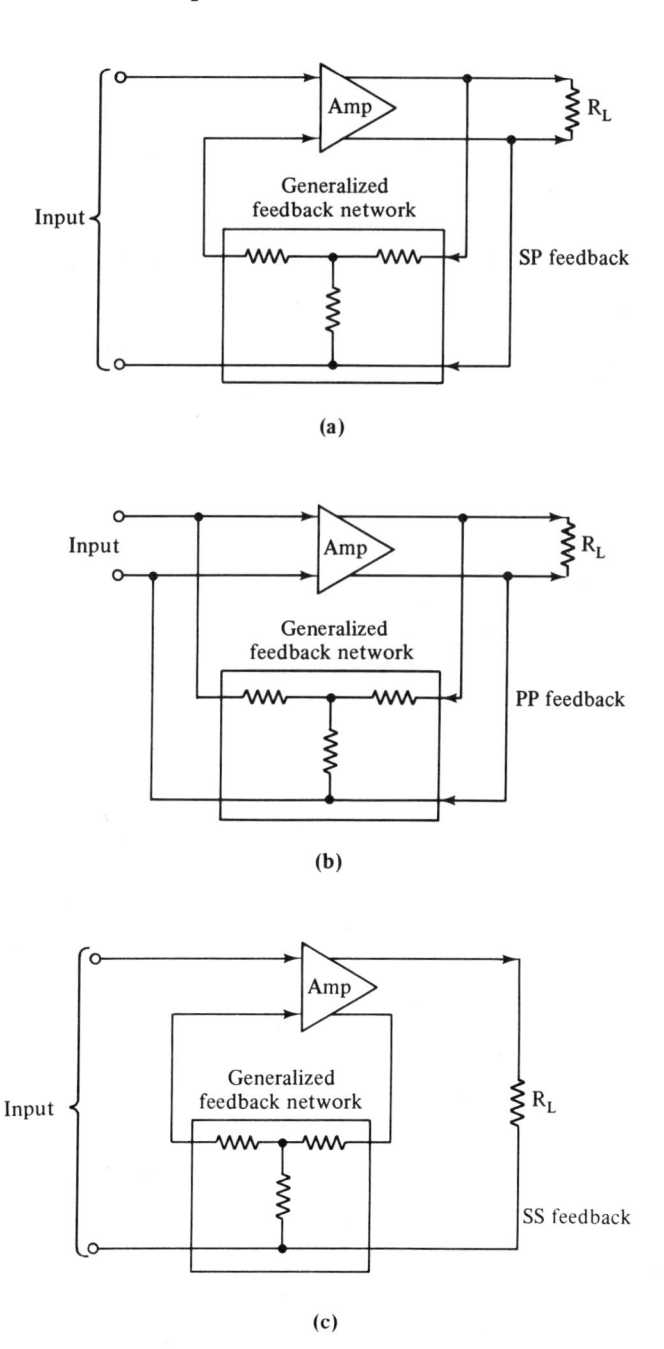

**(c)**

**Figure 3-24.** Four prototype negative-feedback arrangements: (a) series-parallel; (b) parallel parallel; (c) series series.

**(d)**

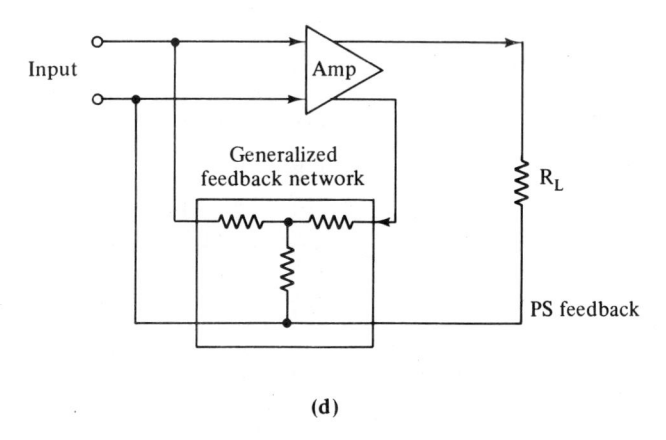

**Figure 3-24.** (*Continued*) (d) parallel series.

are depicted in Fig. 3-25; these are commonly called the voltage-feedback and the current-feedback networks. Here, voltage feedback is provided by a resistor connected from collector to base of the transistor. This is an example of parallel–parallel feedback. Next, current feedback is provided by a resistor connected in series with the emitter lead. Observe that voltage feedback decreases the input resistance and decreases the output resistance of the stage. On the other hand, current feedback increases the input resistance and increases the output resistance of the stage. In either case, the negative-feedback loop functions by opposing the source signal voltage by a certain fraction of the output signal voltage (see Table 3-2 on page 85–87).

This fraction of the output signal voltage that is mixed in phase opposition with the source signal voltage is often termed $B$. Also, the amplification of the stage without negative feedback is often termed $A$. It can be shown that *the reduction in harmonic distortion* that results from the use of negative feedback is stated by the equation

$$D_n = \frac{D_0}{1 + AB}$$

*where*

$D_n$ = percentage of distortion with negative feedback

$D_0$ = percentage of distortion without negative feedback

$A$ and $B$ are defined as noted above

## 3-5 Negative-Feedback Principles

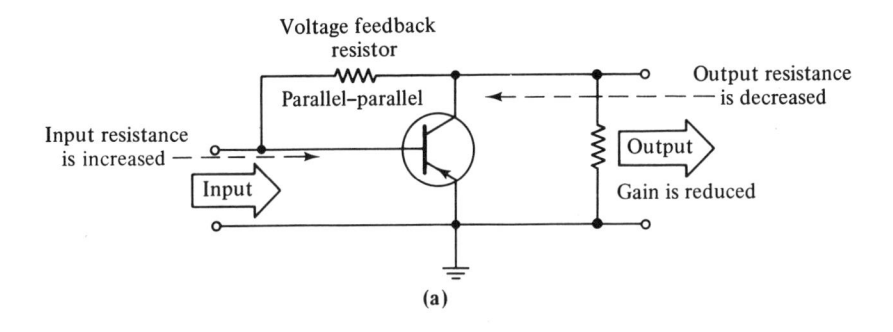

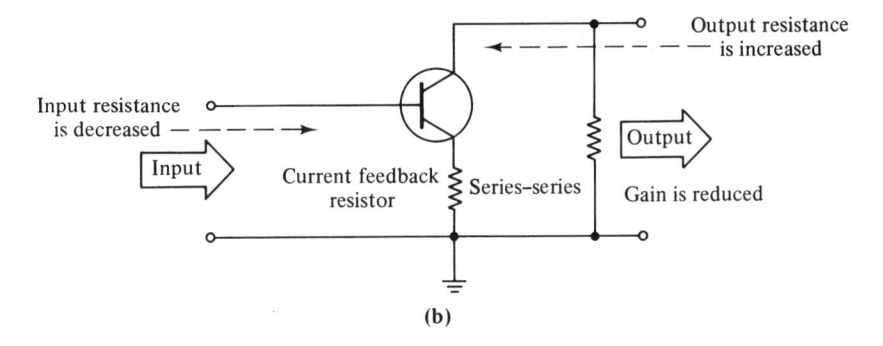

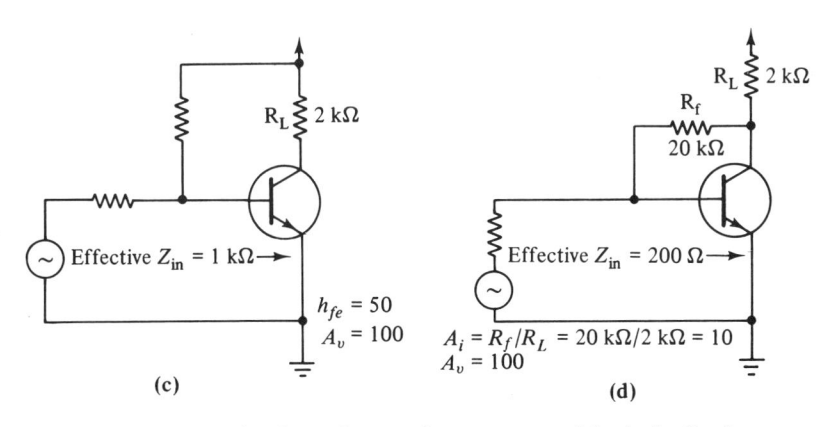

**Figure 3-25.** Skeleton circuits and general parameters of basic feedback arrangements: (a) voltage feedback, also termed shunt feedback or parallel–parallel feedback; (b) current feedback, also termed series feedback or series–series feedback; (c) input resistance and voltage gain of transistor stage with no negative feedback; (d) input resistance, current gain, and voltage gain of stage with shunt feedback.

(a) Voltage feedback resistor, Parallel–parallel, Input resistance is increased, Output resistance is decreased, Gain is reduced

(b) Current feedback resistor, Series–series, Input resistance is decreased, Output resistance is increased, Gain is reduced

(c) Effective $Z_{in} = 1\ k\Omega$, $R_L = 2\ k\Omega$, $h_{fe} = 50$, $A_v = 100$

(d) Effective $Z_{in} = 200\ \Omega$, $R_L = 2\ k\Omega$, $R_f = 20\ k\Omega$, $A_i = R_f/R_L = 20\ k\Omega/2\ k\Omega = 10$, $A_v = 100$

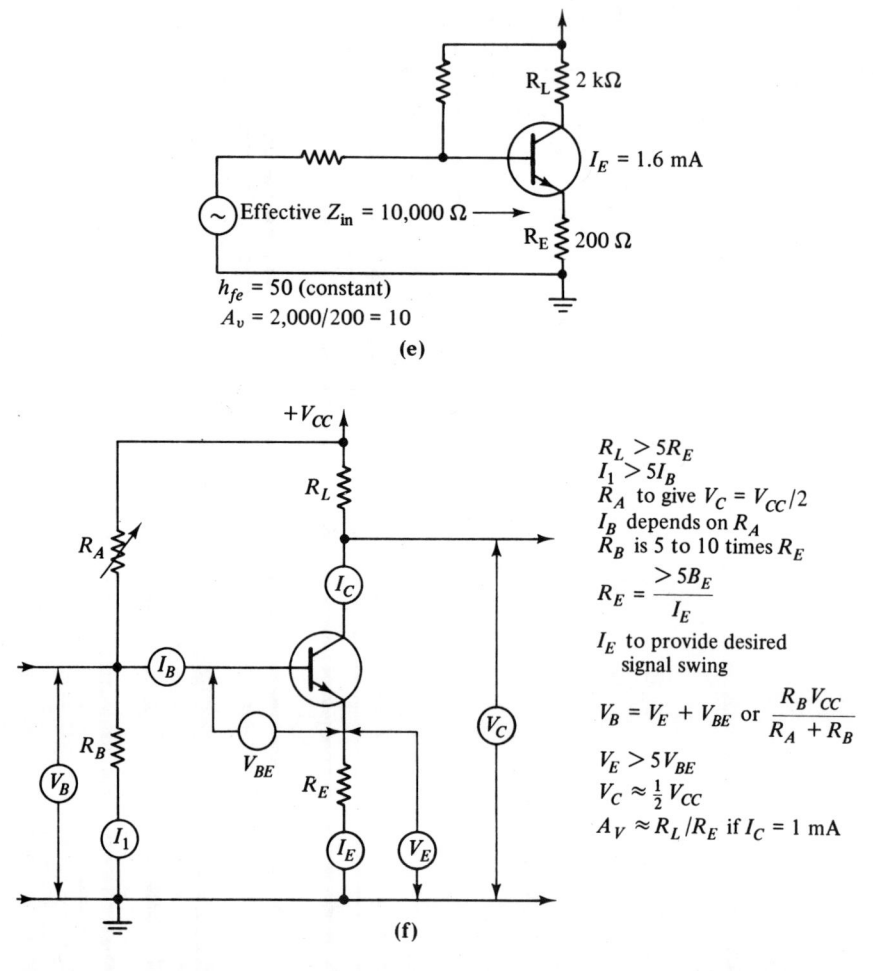

**Figure 3-25.** (*Continued*) (e) input resistance and voltage gain of stage with series feedback; (f) design parameters for common-emitter stage with series feedback.

Diagram (e) labels:
- $R_L = 2\ k\Omega$
- $I_E = 1.6\ mA$
- Effective $Z_{in} = 10{,}000\ \Omega$
- $R_E = 200\ \Omega$
- $h_{fe} = 50$ (constant)
- $A_v = 2{,}000/200 = 10$

Diagram (f) design parameters:
$$R_L > 5R_E$$
$$I_1 > 5I_B$$
$R_A$ to give $V_C = V_{CC}/2$
$I_B$ depends on $R_A$
$R_B$ is 5 to 10 times $R_E$
$$R_E = \frac{>5B_E}{I_E}$$
$I_E$ to provide desired signal swing
$$V_B = V_E + V_{BE} \text{ or } \frac{R_B V_{CC}}{R_A + R_B}$$
$$V_E > 5 V_{BE}$$
$$V_C \approx \tfrac{1}{2} V_{CC}$$
$$A_V \approx R_L / R_E \text{ if } I_C = 1\ mA$$

Negative feedback also improves the frequency response of an amplifier, as depicted in Fig. 3-36. That is, a substantial amount of negative feedback serves to remove humps and sags, and to extend both the high- and low-frequency response of the frequency-response curve. Suppose that an audio-frequency transistor has a beta cutoff frequency of 10 kHz. In turn, the associated amplifier, without negative feedback, will have a frequency response similar to that shown by the solid curve in Fig. 3-27. On

## 3-5 Negative-Feedback Principles

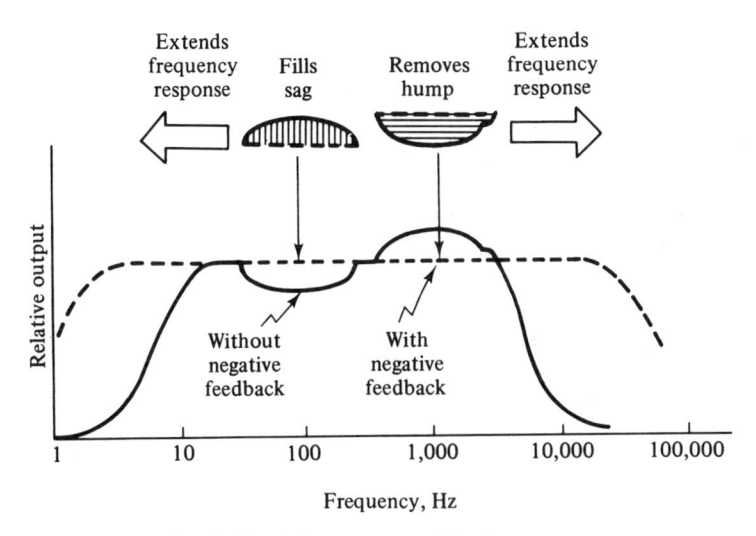

**Figure 3-26.** Negative feedback improves amplifier frequency response.

the other hand, if the amplifier employs 20 dB of negative feedback, the frequency response of the amplifier is improved, as indicated by the dashed curve in the diagram. In other words, a tradeoff of 20 dB in gain provides a 2.7-dB improvement in frequency response at the 3-dB cutoff point.

In addition to the advantages of negative feedback that have been noted, another benefit is relaxation of the tolerance on the load-resistance value. Refer to Fig. 3-22. Suppose that no negative feedback is utilized, and that the value of $R_L$ is increased 20 percent from the indicated value. From a practical point of view, the output voltage will increase 20 percent in turn. That is, the output level will be increased from 2 to 2.4 V, approximately. However, if 20 dB of negative feedback is employed, and the value of $R_L$ is increased 20 percent, the output level will be increased by only 1.5 percent to a level of 2.03 V, approximately. If less negative feedback is used, the degree of output stabilization will be less; but if more negative feedback is utilized, the degree of output stabilization will be greater. In any event, the audio circuit designer can easily calculate the extent to which a stipulated amount of negative feedback will reduce the amplifier output variation as a result of load tolerances.

Still another benefit of negative feedback is its effective stabilization of transistor beta value under conditions of temperature variation. As the operating temperature is increased, a transistor basically responds as follows:

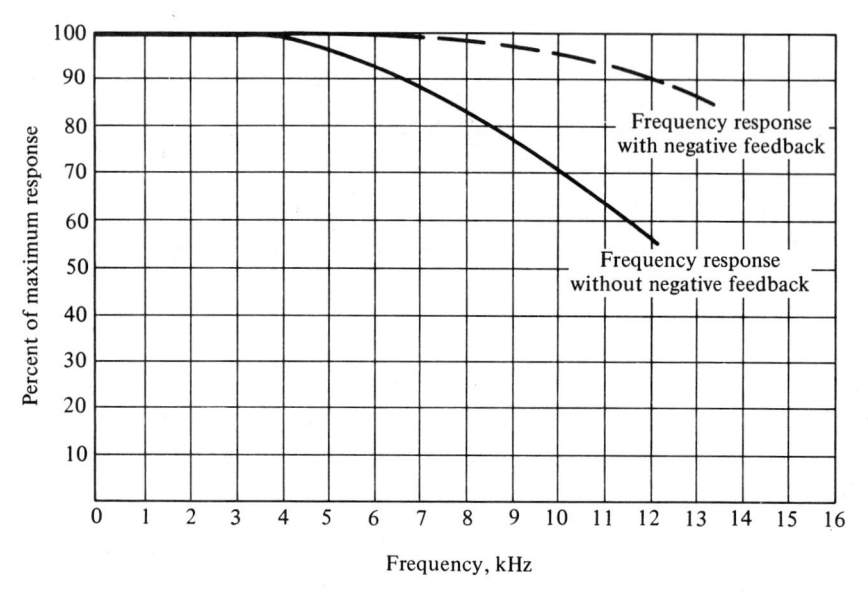

**Figure 3-27.** Improvement of frequency response provided by negative feedback.

1. At collector-current levels above a few milliamperes, the gain (alpha or beta value) of a transistor decreases at a greater rate as the temperature increases.
2. With all other things being equal, the power dissipation of a transistor increases as its operating temperature increases.
3. Collector leakage current doubles for each 8 to 10°C increase in operating temperature.
4. To avoid possible damage, a small-signal transistor should be derated for operating temperatures above 25°C. This derating factor is approximately 1 to 10 mW/°C.
5. Limitation of the maximum junction temperature will avoid excessive leakage-current flow. Bias variation is also reduced by limiting the maximum junction temperature. Both germanium and silicon transistors have a negative temperature coefficient of 2 mV/°C. Temperature changes have less effect on collector-current variation for large values of collector-current flow. Variation of saturation current versus junction temperature is exemplified in Fig. 3-28. Variation of collector current versus junction temperature is shown in Fig. 3-29. The diagram shows that, to maintain a collector-current value of 2 mA as the temperature increases from $x$ to $y$, the base–emitter bias voltage must be reduced from 200 to 150 mV.

## 3-5 Negative-Feedback Principles

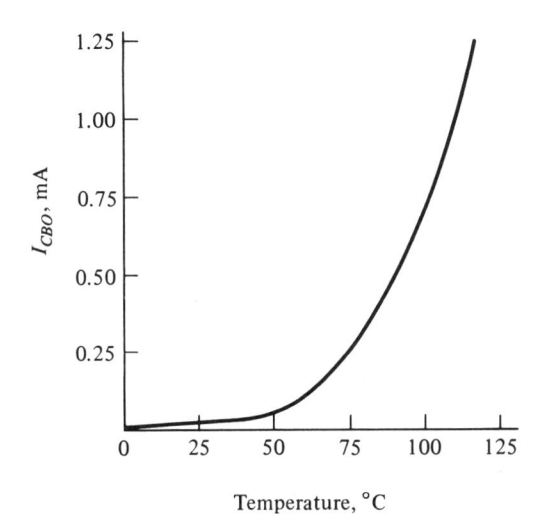

**Figure 3-28.** Variation of saturation current with junction temperature.

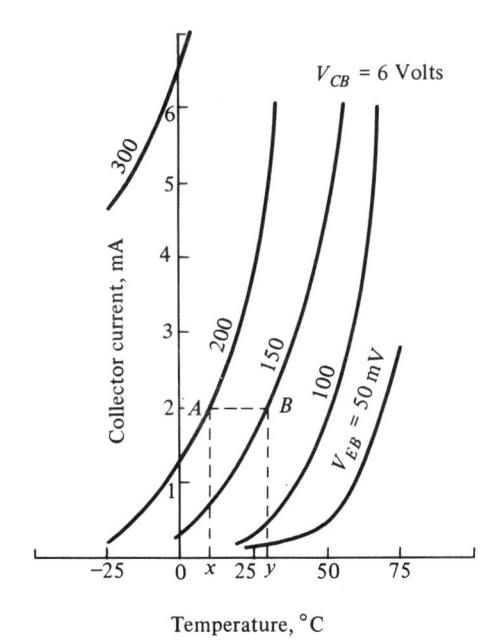

**Figure 3-29.** Variation of collector current with transistor temperature.

As an illustration of the effect of temperature on the transistor transfer characteristic, consider the relations shown in Fig. 3-30. In Fig. 3-30a, are the constant base-current curves vs. collector current for normal tem-

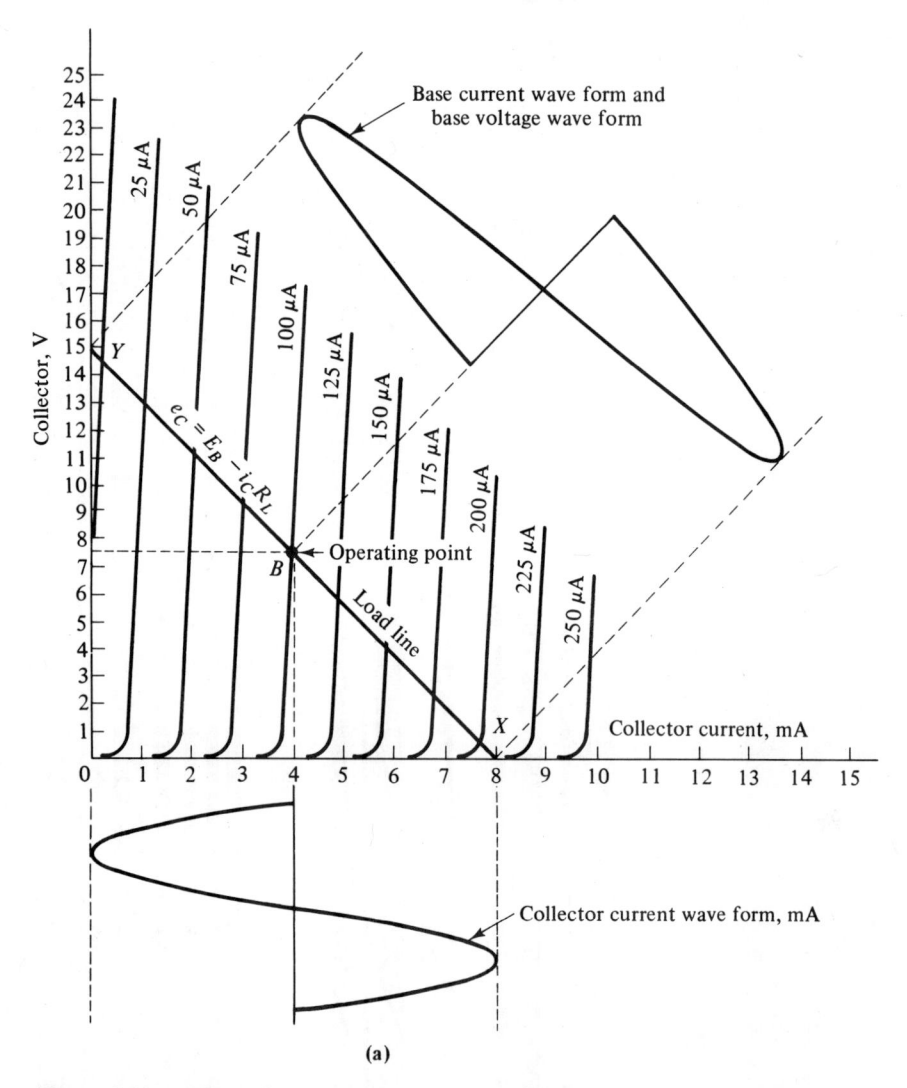

**Figure 3-30.** Effect of heat on transistor transfer characteristic: (a) normal temperature.

perature operation. In Fig. 3-30b, we observe the curves for the same transistor at an increased operating temperature. Note that the effect of heat is to move the curves to the right along the $X$ axis, and (in some situations) to increase the spacing between the curves. In turn, the operating point $B$ is shifted on the load line toward point $X$, thereby reducing the collector voltage. The normal collector voltage for class A operation, at point $B$ in Fig. 3-30a, is 7.5 V. However, increased temperature shifts

## 3-5 Negative-Feedback Principles

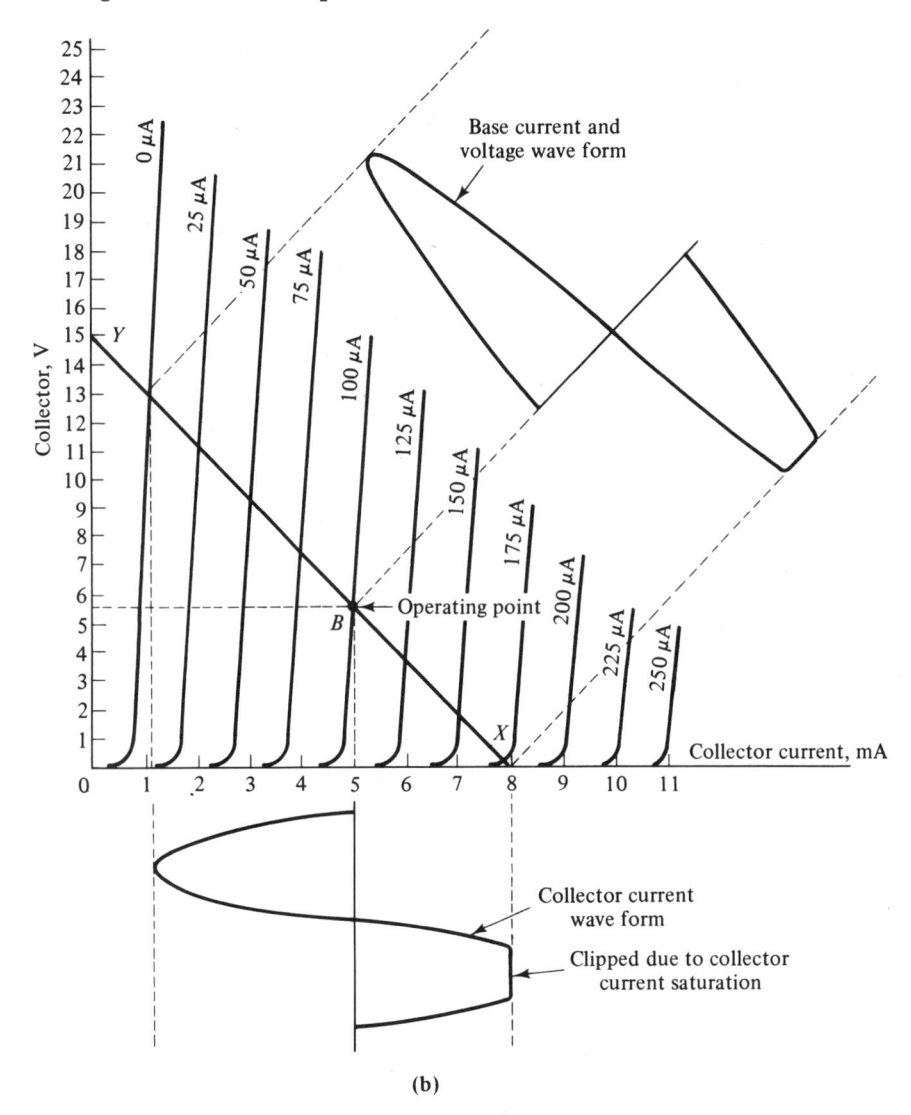

**Figure 3-30.** (*Continued*) (b) increased temperature.

point $B$ to a point indicated in Fig. 3-30b that corresponds to a collector voltage of 5.5 V. From the viewpoint of the audio circuit designer, the original input signal now overloads the amplifier, and peak clipping occurs as a result of collector-current saturation. It will be shown that dc negative feedback can be utilized to avoid this difficulty.

Next consider the stabilization of beta value that is provided by negative feedback under conditions of temperature variation. It has been noted that a semiconductor junction has both ac and dc resistance values. Accordingly, a transistor has ac and dc beta values. The former value is usually symbolized by $h_{fe}$ and the latter by $h_{FE}$. Variations in ac and dc beta values for a typical germanium transistor are graphed in Fig. 3-31. Observe that the dc beta is more responsive to temperature increase than is the ac beta value. Here we will note only the stabilization of ac beta by means of negative feedback; stabilization of dc beta will be detailed subsequently under bias-circuit design considerations. With reference to Fig. 3-31, the value of $h_{fe}$ varies by approximately 3 to 1 over the temperature range from $-55$ to $+85°C$. This variation is equivalent to a $\pm 50$ percent tolerance on the beta value. However, if 20 dB of negative feedback is employed, as in the example of Fig. 3-22, the effective tolerance on the beta value due to temperature variation is reduced to $\pm 10$ percent.

Consider the example of dc negative feedback shown in Fig. 3-32. This is an audio mixer configuration. It is basically an audio preamplifier with four inputs that can be operated simultaneously, if desired. A junction-field-effect transistor (JFET) is utilized in the input stage; it operates in the common-source mode with a bypassed source resistor. A bipolar transistor is used in the output stage; it operates in the emitter-follower mode with 100 percent current feedback across R11. The frequency response of this audio mixer is approximately $\pm 3$ dB from 20 Hz to 100 kHz. Resistor R10 develops source bias for Q1 and provides dc negative feedback. Because R10 is bypassed, it does not develop any ac negative

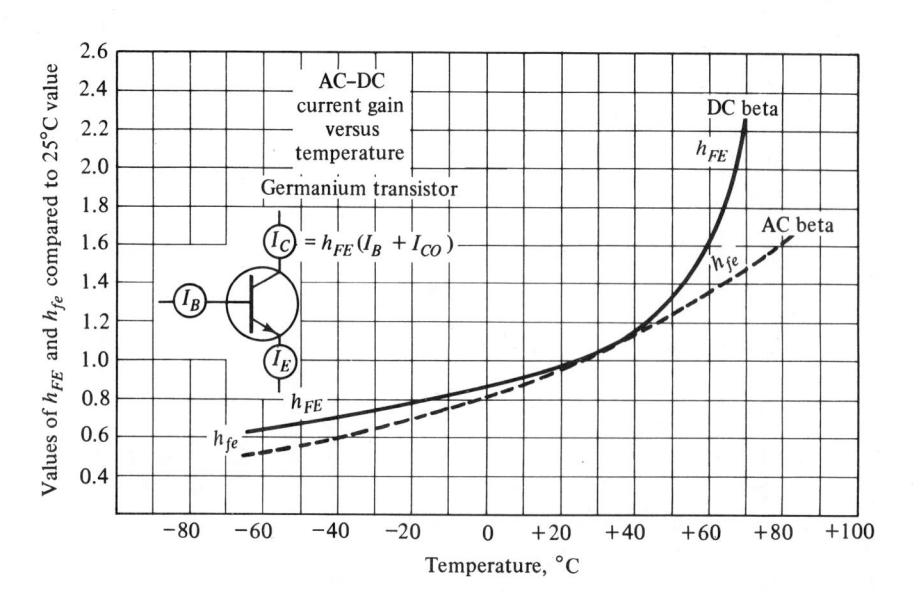

**Figure 3-31.** Variation of ac beta and dc beta values versus temperature.

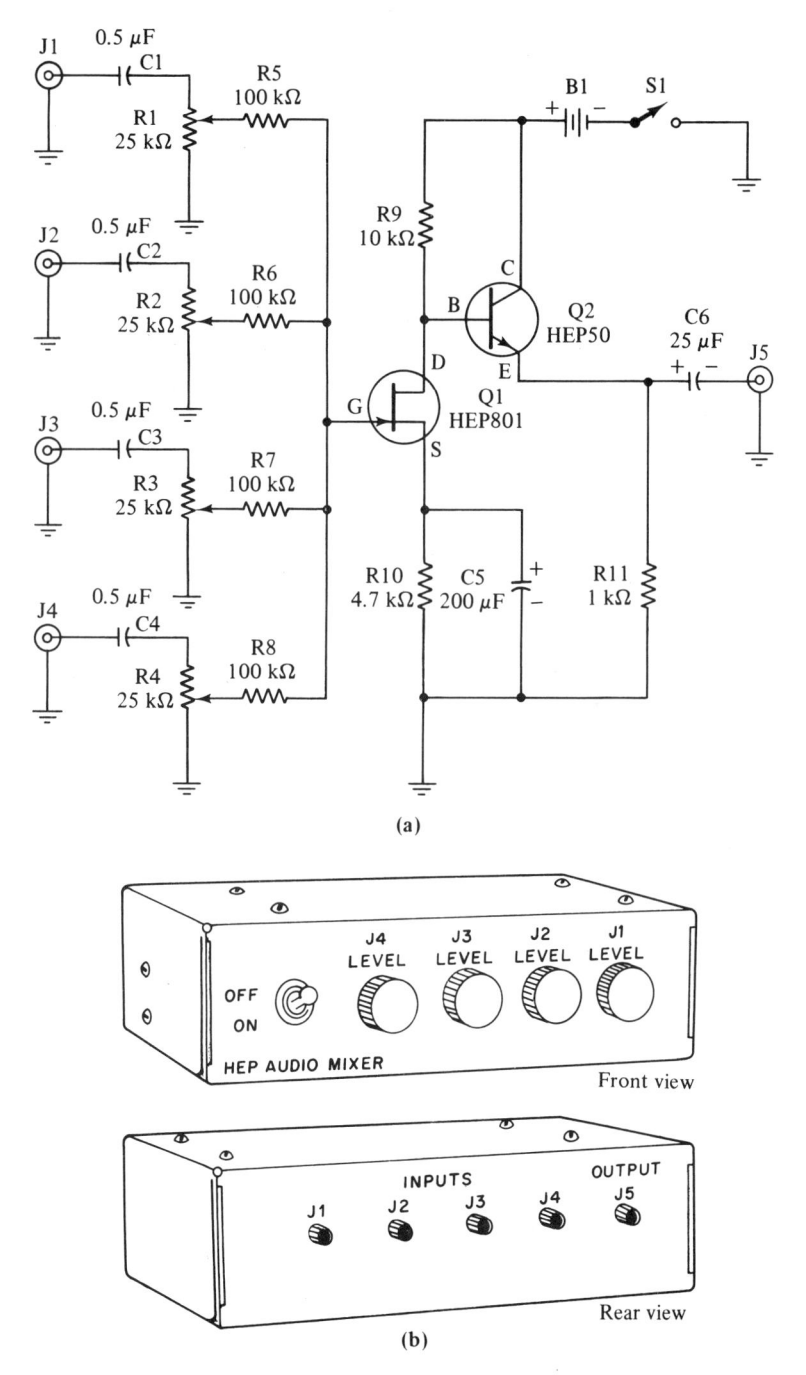

**Figure 3-32.** Basic audio mixer: (a) configuration; (b) appearance. (*Courtesy, Motorola*)

103

feedback. On the other hand, R11 develops both ac and dc negative feedback. The voltage gain of this audio mixer is from 5 to 10 dB, depending upon FET tolerances. Maximum output signal level is 1 V rms.

If desired, C5 may be omitted and R10 thus operated in the unbypassed mode. This design variation increases the fidelity of amplifier operation to some extent at the expense of stage gain. The effect of degeneration due to an unbypassed source resistor is to reduce the stage gain according to the following equation:

$$\frac{A_{fb}}{A} \approx \frac{1}{1 + AB}$$

where

$A_{fb}$ = stage amplification with degeneration

$A$ = stage amplification without degeneration

$B$ = ratio of the source resistance to the sum of the source and drain resistances

## 3-6 Multistage Negative-Feedback Relations

Single-stage negative-feedback circuitry is in very wide use. It is common practice to employ unbypassed emitter resistors to obtain signal degeneration. Voltage feedback from collector to base of the same transistor is also utilized in many amplifier configurations. Negative feedback over two stages is in less extensive use, although it is rather common. As an illustration, voltage feedback can be applied from the collector of the second-stage transistor to the emitter of the first-stage transistor. On the other hand, three-stage negative feedback is rarely used, because of its inherent instability hazard. In theory, and in the first analysis, it may appear quite feasible to apply negative feedback from the collector of the third-stage transistor to the base of the first-stage transistor. However, inherent phase shifts in the cutoff region impose a serious problem, as explained next.

A single stage of amplification has an amplitude-phase characteristic similar to that shown in Fig. 3-33a. Observe that zero phase shift occurs only at the midband frequency. As the frequency characteristic progresses through the cutoff regions, at the high and at the low end of its range, the phase characteristic varies rapidly and approaches a limit of either 90 degrees leading or 90 degrees lagging. When two stages are connected in cascade, the phase characteristic approaches a limit of 180-degree shift at the extreme ends of its frequency range. Or if three stages are connected

## 3-6 Multistage Negative-Feedback Relations

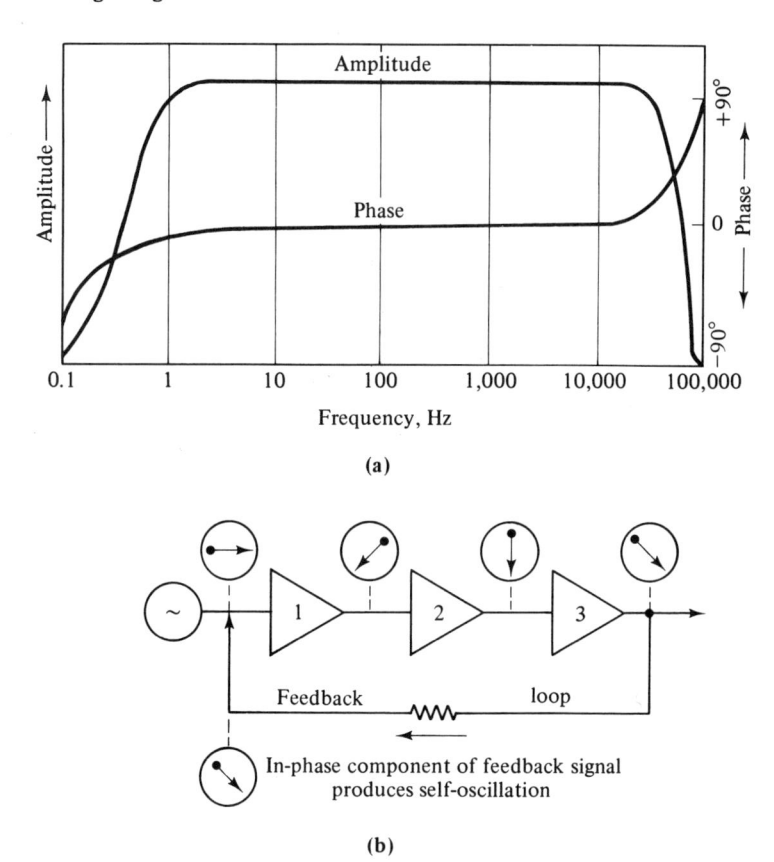

**Figure 3-33.** Amplitude-phase characteristics of an amplifier stage: (a) phase versus amplitude response; (b) phase shifts are cumulative.

in cascade, the phase characteristic approaches a limit of 270-degree shift as a limit. If the overall phase shift is such that the negative feedback voltage has an in-phase component with respect to the source voltage at some frequency, the amplifier will go into self-oscillation at that frequency. In other words, the feedback has become positive instead of negative, owing to cumulative phase shift.

With reference to Fig. 3-33b, the three stages operate in the CE mode; thus there is a 180-degree phase shift from base to collector at the midband frequency. However, the phase shift departs from 180 degrees in the cutoff regions. Thus, for the operating condition exemplified in the diagram, the negative-feedback voltage is applied at the input of the first stage with a 45-degree in-phase component. Consequently, the amplifier supplies its own input and breaks into sustained self-oscillation. This malfunction is more likely to occur at the low-frequency end of the band than

Gain = $\dfrac{A}{1 + AB}$

$A$ = gain without negative feedback
$B$ = portion of output signal that is fed back

Locus of $AB$ versus frequency

Nyquist curve

Real axis

(1, 0)

Imaginary axis

Amplifier is stable if point (1, 0) is not enclosed by the Nyquist curve

(c)

Amplification

Input

Part of input is canceled

Output

Part of output is fed back

Feedback loop

(d)

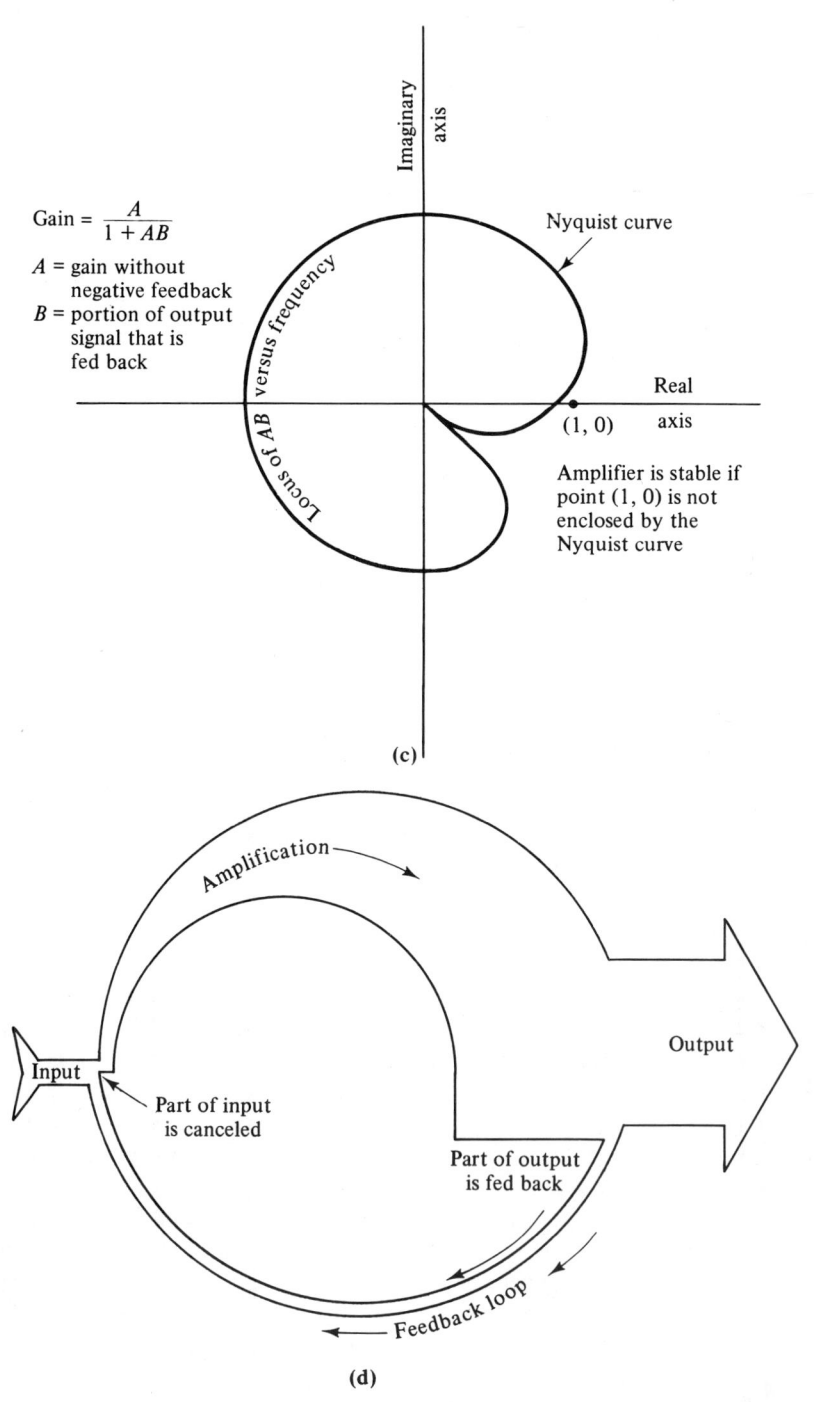

**Figure 3-33.** (*Continued*) (c) Nyquist plot shows if a feedback arrangement is stable; (d) negative-feedback circuit action.

### 3-6 Multistage Negative-Feedback Relations

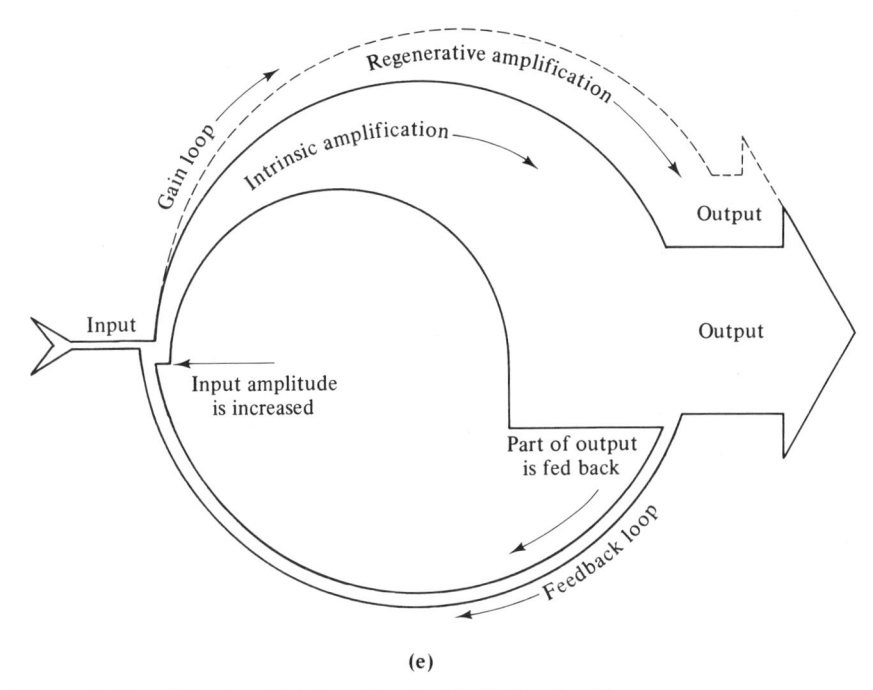

**Figure 3-33.** (*Continued*) (e) positive-feedback circuit action.

at the high-frequency end. In other words, self-oscillation will occur at the frequency for which the system has its highest $Q$ value. Inasmuch as an audio transistor generally has more low- than high-frequency gain, low-frequency self-oscillation will ordinarily ensue.

Consider a prototype model of an amplifier that has been designed with carefully chosen component and device values, so that multistage negative feedback is stabilized. In turn, there is no assurance that the design will remain stable when transistors with higher beta values, or with higher or lower beta cutoff frequencies are utilized in production. Similarly, the configuration may become unstable when resistors and capacitors with perhaps ±20 percent tolerance are used on the assembly line. It follows that, whenever the audio circuit designer employs negative feedback over more than one stage, he should make an intensive worst-case analysis of potential instability due to cumulative phase shifts. Also, worst-case analysis must take into consideration any anticipated temperature variations and supply-voltage variations. If there is a possibility that the amplifier may be used to drive an inductive or a capacitive load, instead of a purely resistive load, this factor should be taken into account. Again, if the amplifier may be used with a generator that has a substantially different value of internal impedance than is used to test the prototype model, this operating parameter should also be checked.

## 3-7 Transistor Parameter Variation

Various transistor characteristics, limitations, and operating factors should be taken into consideration by the audio circuit designer to make a thorough worst-case analysis:

1. *Deterioration with time.* Conservative ratings of transistor parameters should be assigned to allow for anticipated deterioration in operating characteristics with time.
2. *Current gain versus temperature.* At collector-current levels above a few milliamperes, the alpha or beta value of a transistor decreases at a greater rate as the temperature increases.
3. *Power dissipation versus temperature.* Circuit design provisions may be required to avoid excessive power dissipation by a transistor at high temperatures. When breadboard worst-case analyses are made, it is good foresight to include a fuse or a current-limiting resistor to avoid accidental damage to the transistor as a result of excessive collector-current flow.
4. *Frequency limitations.* Frequency cutoff limits in common-emitter or common-collector configurations depend on forward current-gain values. Collector junction capacitance is a contributing factor in limitation of high-frequency response.
5. *Collector leakage currents.* Collector leakage currents will double for each 8 to 10°C increase in operating temperature. In the CE configuration, $I_{CE}$ can vary from $I_{CBO}$ to $h_{FE} \times I_{CBO}$. It is good practice to minimize the circuit resistance between base and emitter, insofar as stage gain is not adversely affected.
6. *Manufacturers' ratings.* Second-source considerations should be taken into account; different manufacturers may publish somewhat different ratings for the same transistor type number.
7. *Power considerations.* A suitable thermal derating factor should be applied for operating temperatures above 25°C. This factor is approximately 1 to 10 mW/°C for small-signal transistors, and 0.25 to 1.5 W/°C for power-type transistors. The thermal resistance value ($R_T$, $\theta_R$) in a manufacturer's rating does not include the thermal resistance of the heat sink. It is good practice to include an emitter swamping resistance, whenever this is practical, to avoid the possibility of thermal runaway.
8. *Temperature considerations.* Excessive leakage currents can be avoided by limiting the maximum junction temperature. Bias variations can be reduced by limiting the minimum junction temperature. Both germanium and silicon transistors have a

negative temperature coefficient of 2 mV/°C. Large values of collector current ($I_c$) assist in reduction of collector-current variation due to temperature changes. Low values of source resistance in driving a base circuit contribute to a low stability factor. When diodes are used for bias stabilization of a transistor, a common heat sink should be provided for the devices. Conservative circuit design observes a minimum $f_{FE}$ value over the operating temperature range. Note that a low stability factor has no significance insofar as the performance of a direct-coupled amplifier is concerned.

9. *Operating voltage precautions.* The audio-circuit designer should never exceed the $V_{CB}$ maximum, $V_{CE}$ maximum, or $V_{BE}$ maximum (reverse breakdown voltage) ratings under any condition of circuit operation. In a push–pull configuration, $V_{CC}$ should be less than half of $V_{CB}$. Avoid any possibility of transients (such as switching transient voltages) that could exceed the maximum ratings. Remember also that the reverse breakdown voltage of a silicon transistor decreases with increasing temperature.

10. *Soldering precautions.* Transistor damage can occur in production operations as a result of excessive heat while the terminals are soldered. In a dip-soldering process during assembly of printed circuits with transistors, the solder temperature should not exceed 250°C for a maximum immersion period of 10 s. The flange of a transistor must not be soldered to a heat sink.

## 3-8 Production Cost Tradeoffs

As a general rule, it will be found that a class A audio amplifier with a rated power output of less than 0.1 W will be least expensive to produce if it is designed with both direct-coupled input and output. On the other hand, it is more difficult to realize good bias stability versus temperature variation than if transformer coupling were employed. Again, an audio amplifier with a rated power output of more than 0.1 W is usually least expensive to produce if it is designed with transformer-coupled output and direct-coupled input. However, the bias stability of this arrangement versus temperature variation is poorer than if input transformer coupling were utilized. In high-fidelity design, transformer coupling has the disadvantage of introducing more harmonic distortion than if direct coupling were used. Another tradeoff that the audio circuit designer must contend with is the comparatively limited frequency range that is inherent in transformer-coupled circuitry.

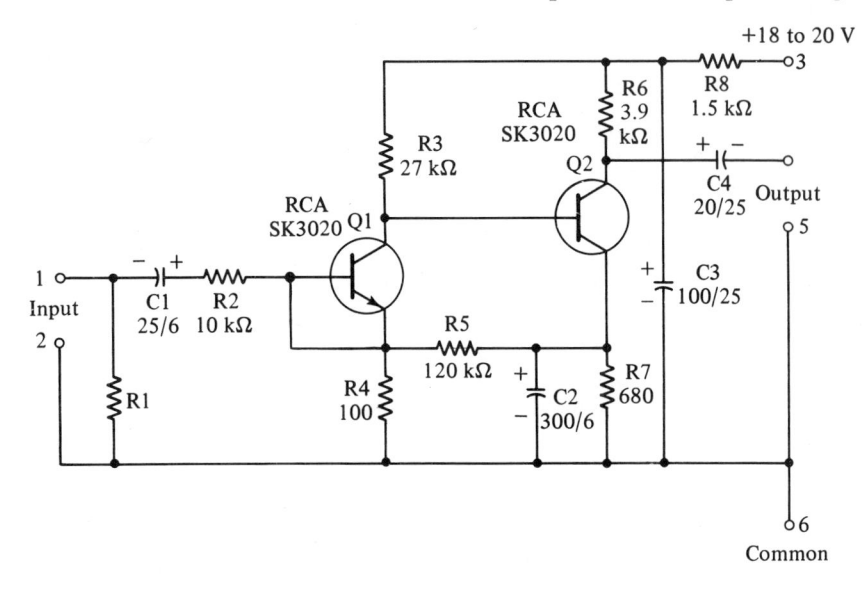

**Figure 3-34.** Preamplifier with direct-coupled input and direct-coupled output circuitry. (*Courtesy, RCA*)

Parts List
C1: 25 µF, 6 V, electrolytic
C2: 300 µF, 6 V, electrolytic
C3: 100 µF, 25 V, electrolytic
C4: 20 µF, 25 V, electrolytic
Q1, Q2: transistor, RCA SK3020
R1: 220 Ω for low-impedance microphone, 270,000 Ω for high-impedance microphone, 1/2 W, 10%
R2: 10,000 Ω, 1/2 W, 10%
R3: 27,000 Ω, 1/2 W, 10%
R4: 100 Ω, 1/2 W, 10%
R5: 120,000 Ω, 1/2 W, 10%
R6: 3,900 Ω, 1/2 W, 10%
R7: 680 Ω, 1/2 W, 10%
R8: 1,500 Ω, 1/2 W, 10%

A well-designed configuration for an all-purpose microphone preamplifier with both direct-coupled input and output is shown in Fig. 3-34. It provides an output level from 0.5 to 1 V from a dynamic microphone, with a frequency response from 20 Hz to 35 kHz. Component tolerances are noted in the parts list. The microphone source impedance may range from 200 Ω to 30 kΩ. Base bias current for Q1 is obtained from the emitter of Q2 via R5. Base bias current for Q2 is obtained through the collector resistor of Q1. Thereby, dc feedback is provided for stabilization of the transistor operating points. The current drain of the amplifier is approximately 2.5 mA, and the voltage gain is about 1,700 times.

If the preamplifier is used with a low-impedance dynamic microphone, the value of R1 should be reduced to 220 Ω. On the other hand, if a microphone with an impedance of 30 kΩ is utilized, the value of R1

## 3-8 Production Cost Tradeoffs

should be increased to 270 kΩ. A drilling template and a component placement diagram for the preamplifier circuit board are depicted in Fig. 3-35. A 2- by 3-in. circuit board is suitable for a single preamplifier circuit; two preamp circuits can be assembled on a 3- by 4-in. circuit board. If the audio circuit designer does not fabricate the assembly on a circuit board, a ground bus should be used to ground the preamplifier to the circuits that follow it at one point only. It is good practice to choose the input of the circuits as a common ground point. The component layout depicted in Fig. 3-35 is designed for printed circuitry.

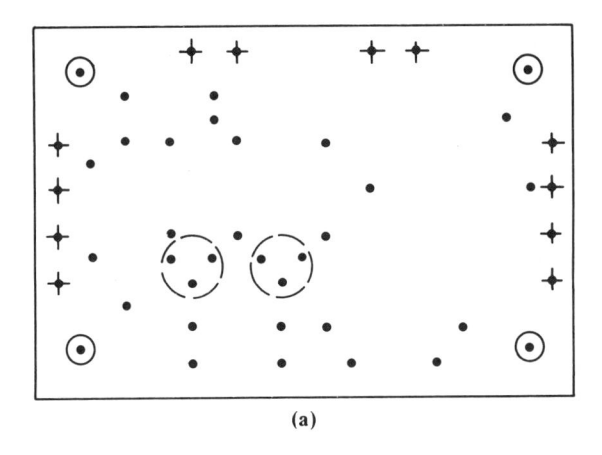

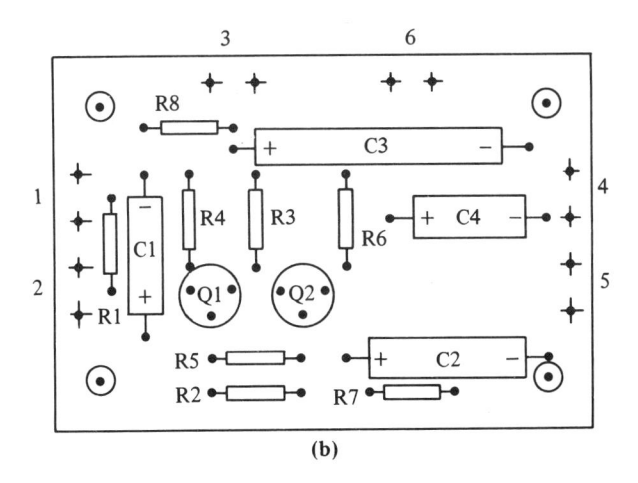

**Figure 3-35.** Circuit board layout for the exemplified preamplifier: (a) drilling template; (b) component placement. (*Courtesy, RCA*)

## 3-9 Notes on Musical and Speech Wave Forms

Vocal and musical tones are reproduced by wave forms that have sharp and high peaks, as exemplified in Fig. 3-36. The music-power rating of an amplifier is defined as the peak power value that can be delivered to the speaker for a very short period of time, with no more harmonic distortion than at maximum rated sine-wave power output. This music-power rating describes the ability of an amplifier to handle sudden peak musical wave forms without distortion. Peak duration in this mode is usually limited by the ability of the filter capacitors in the power supply to deliver the peak current demand. Music-power ratings are determined with pulse signals; a typical pulse generator is illustrated in Fig. 3-37. The output pulse wave form is observed on an oscilloscope screen. A pulse width of 1 millisecond (ms) is utilized with a repetition rate of 100 pulses per second.

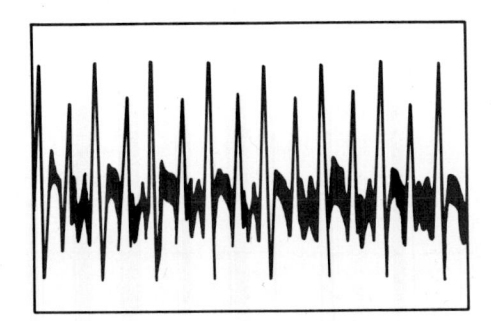

**Figure 3-36.** Vocal and musical wave forms have high sharp peaks.

**Figure 3-37.** Pulse generator. (*Courtesy, CSC*)

## 3-10  Saturation Currents

There are two basic forms of saturation current that should not be confused with each other. One form is defined as the heavy current that flows between the base and the collector of a bipolar transistor when an increase in forward bias causes no further increase in the collector current. If there is 1 kΩ or more of series resistance in the collector circuit, the heavy current flow in the saturated state of operation will result in a very low value of collector-emitter voltage. Both transistor junctions will be forward biased while the transistor is in saturation, and nearly all the $V_{CC}$ voltage will be dropped across the series resistance in the collector circuit. The saturation region is depicted in Fig. 3-38.

On the other hand, the other form of saturation current is very small; it is the reverse current that flows across the collector–base junction while the junction is reverse biased. Although this form of saturation current is very small, it is very important in transistor operation over appreciable temperature ranges. In other words, this saturation current can cause serious difficulties in amplifier operation at comparatively high ambient temperatures unless the bias circuits are properly designed. This requirement is detailed in the following chapter.

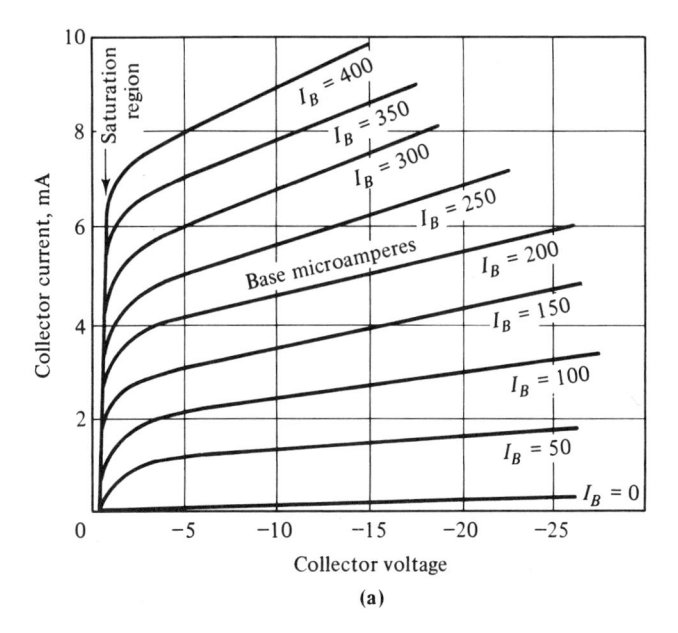

**Figure 3-38.** Saturation region of a bipolar transistor: (a) conventional collector family diagram.

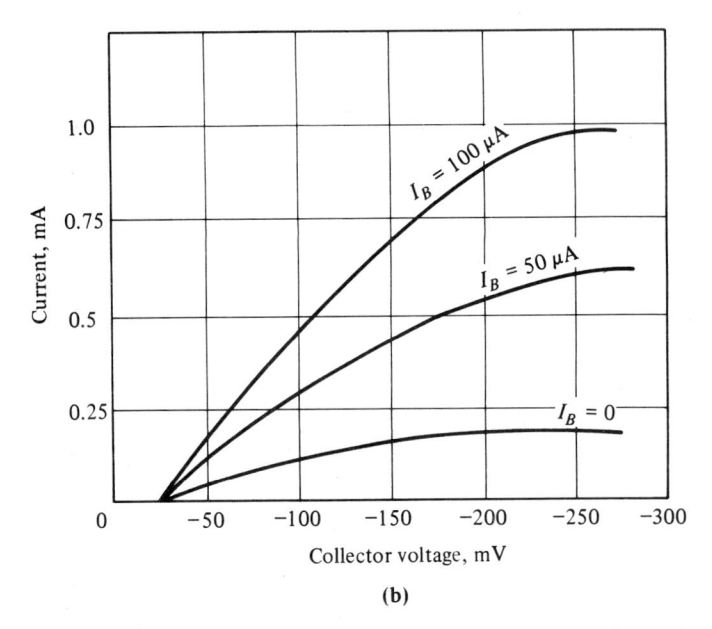

**Figure 3-38.** (*Continued*) (b) expanded section, showing the saturation region.

## 3-11 Bootstrapping Method of Feedback

In addition to negative feedback, positive feedback is also employed in certain types of amplifier configurations. This technique is often termed *bootstrapping;* it is utilized to increase the input impedance of an amplifier, or to equalize stage gains in specialized types of push–pull circuitry. Consider the Darlington connection, shown in Fig. 3-39a. It is used, for example, in solid-state applications to provide a high input impedance. Next observe the Darlington arrangement with bias circuitry depicted in Fig. 3-39b. It is evident that the input impedance of the stage is limited by the value of R3 in parallel with R2 + R1. However, if a bootstrap capacitor $C_F$ and a bootstrap resistor $R_F$ are connected from the emitter of Q1 to the junction of R1 and R2, sufficient feedback is provided so that the base circuit of Q1 draws very little signal current. In accordance with Ohm's law, this reduction in current is equivalent to a large increase of Q1's input impedance. Since the bootstrap loop bleeds signal power from the output of Q1, the drive level to Q2 is reduced accordingly, and the gain of Q1 is effectively reduced. This tradeoff in gain is accepted for a corresponding increase in input impedance.

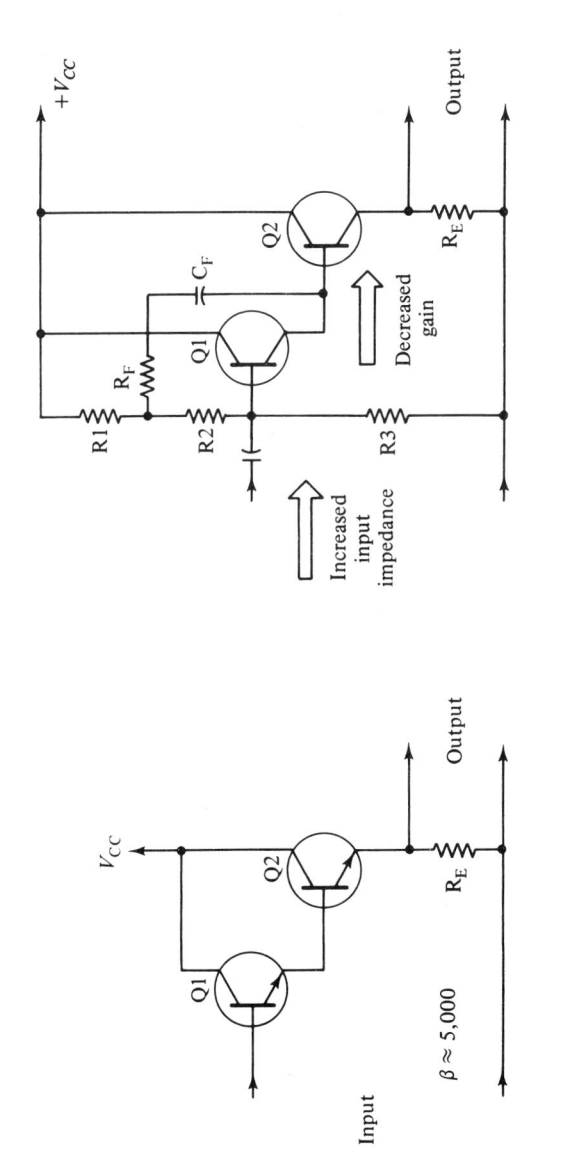

**Figure 3-39.** Example of feedback with a bootstrap connection: (a) basic Darlington pair; (b) Darlington pair with bias circuit and positive-feedback loop (bootstrap connection).

## 3-12 Frequency Response of Cascaded Stages

When conventional audio-amplifier stages are cascaded without negative feedback and without any frequency-compensating networks, the bandwidth of the amplifier becomes less as the number of stages is increased. Also, the rolloff at the high-frequency end of the response curve approaches a limiting shape that is called the *Gaussian response,* as depicted in Fig. 3-40. In theory, an infinite number of stages would have to be cascaded to obtain a true Gaussian response. In practice, this limiting curve shape is approached rather rapidly. This curve is basically the same as the rolloff of a standard probability curve. It is of interest to the circuit designer in some situations, because a Gaussian response provides least transient distortion and fastest rise for a given bandwidth.

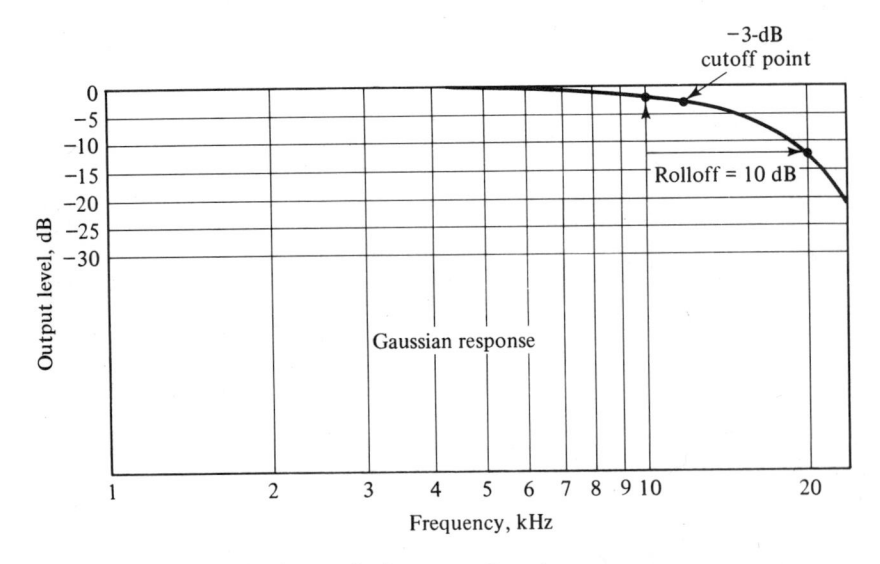

**Figure 3-40.** Example of an audio-frequency Gaussian response.

chapter four

# AMPLIFIER BIAS STABILIZATION METHODS

## 4-1 General Considerations

Most of the voltages in an audio-amplifier configuration have comparatively wide tolerances, with one exception. In other words, transistor bias voltages are subject to unusually tight tolerances. One basic reason for the critical aspect of bias voltages is the nonlinearity of the transfer characteristic for a bipolar transistor, as exemplified in Fig. 4-1. In other words, the collector current increases out of proportion to the base–emitter bias voltage; if the bias voltage increases by a small amount, the collector current increases by a large amount. Accordingly, the audio circuit designer must utilize bias circuitry that maintains a tight tolerance on the bias voltage, regardless of temperature variation, supply-voltage variation, and comparatively wide tolerances on transistor characteristics. We will find that simple amplifier arrangements, such as shown in Fig. 4-2, have poor bias stability.

  Bias corresponding to the operating or quiescent point for a transistor is established by first assigning particular fixed values of collector voltage and emitter current. Then the required value of base–emitter bias voltage is determined for this value of emitter current. Since the input and output ports of a bipolar transistor are not entirely independent of each other, the base current that flows depends both upon the value of the base–emitter bias voltage and also upon the collector–emitter voltage, as exemplified in Fig. 4-3. The quiescent point is calculated or determined with no applied signal, and it is specified in terms of dc voltage and current

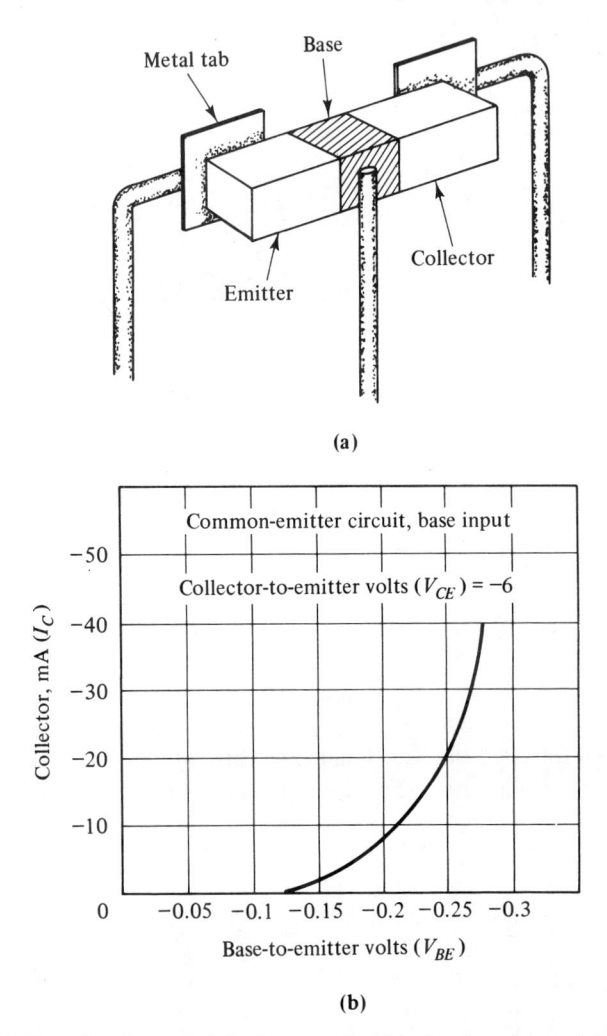

**Figure 4-1.** Transfer characteristic for a typical bipolar transistor: (a) basic structure of a bipolar transistor; (b) bias-voltage/collector-current relation.

values. Basic bias requirements are determined for operation at room temperature.

Reliable operation of a transistor over a wide temperature range requires some means of stabilization for the base voltage and base current values. Thus compensating circuits are employed to obtain bias stability under conditions of variation in reverse-bias collector current (saturation current) and variation in emitter–base junction resistance. It is essential to recognize the variation in saturation current that occurs with temperature change of the base–collector junction, as depicted in Fig. 4-4. Note

## 4-1 General Considerations

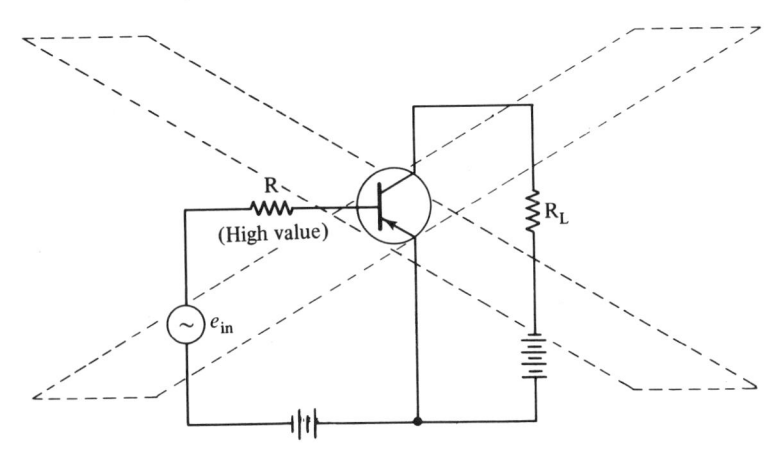

**Figure 4-2.** Simple amplifier arrangement that has poor bias stability.

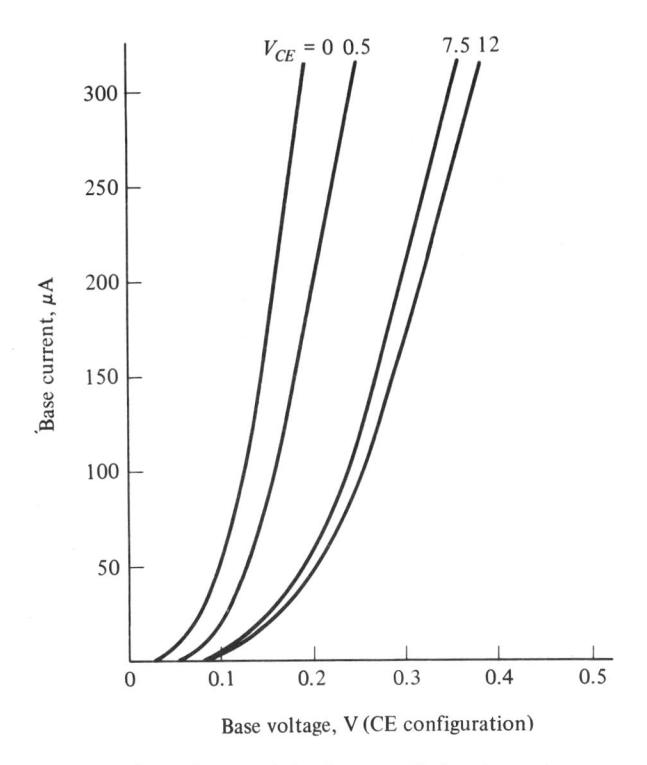

**Figure 4-3.** Base–emitter characteristics for a small-signal transistor.

that the saturation current is almost zero at 10°C and then increases out of proportion to temperature, so that its value is considerably greater than

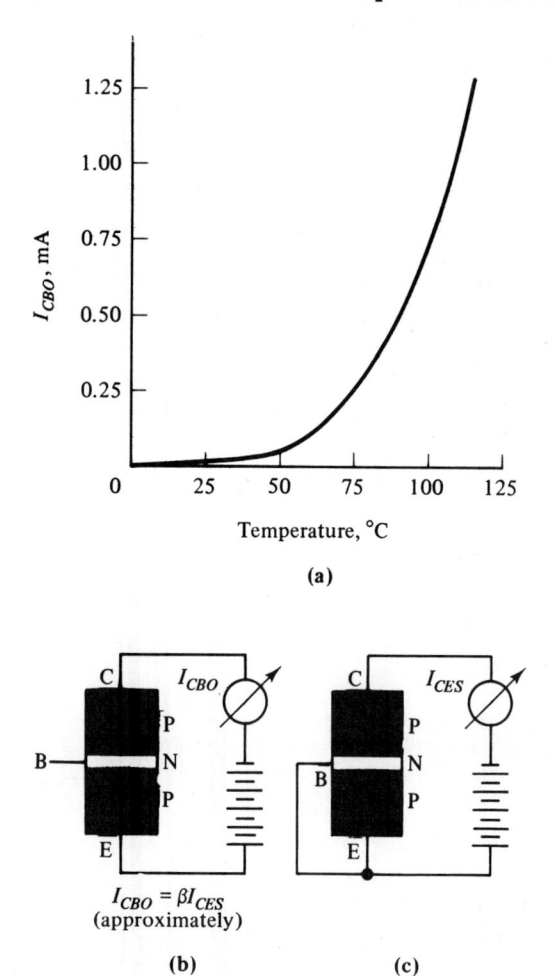

**Figure 4-4.** Variation of saturation current $I_{CBO}$ with junction temperature: (a) $I_{CBO}$ value versus temperature for a germanium transistor; (b) measurement of $I_{\text{CBO}}$; (c) measurement of $I_{\text{CES}}$.

1 mA at 125°C. It is apparent that at temperatures below 10°C, saturation current has virtually no effect on bias conditions. Observe that Fig. 4-3 shows the variation in base-current flow versus operating potentials, whereas Fig. 4-4 depicts the variation in saturation current (reverse collector current) versus temperature.

It is evident that saturation current increases very rapidly at higher temperatures; we will show that this increase in saturation current causes a rapid increase in emitter current (in simple circuits), or that an increase in saturation current causes a rapid increase in collector current if it is uncon-

## 4-1 General Considerations

trolled. In turn, bias circuitry must be designed for effective stabilization of collector current under conditions of temperature variation with its related variation in saturation current. An added benefit that will appear in conjunction with bias stabilization of an amplifier circuit is the fact that it also makes the amplifier action relatively independent of tolerances on saturation current in replacement transistors. $I_{CBO}$, commonly called $I_{CO}$, is the dc collector current that flows when a specified voltage $V_{CBO}$ is applied from collector to base, the emitter terminal being left open (unconnected). The polarity of the applied voltage is such that the collector–base junction is biased in a reverse direction. Again, $I_{CES}$ is the dc collector current that flows when a specified voltage is applied from collector to emitter, the base being short-circuited to the emitter. The polarity of the applied voltage is such that the collector–base junction is biased in a reverse direction.

Saturation current, in the case of an NPN transistor, consists of hole flow from the collector into the base. Refer to Figs. 4-4b and 4-4c; the value of $I_{CBO}$ is approximately beta times the value of $I_{CES}$. If resistors connected to the base have a high value, the saturation-current path can be represented by an open-base condition. In turn, holes can accumulate in the base region. This charge accumulation causes an increase of emitter current (electrons) into the base, and increases the collector current. This circuit action dissipates power and increases the temperature of the collector–base junction, thereby causing the saturation-current value to increase further. This device action will continue until the amplifier stage distorts seriously; in fact, the "snowballing" action (thermal runaway) attains such a high level that the transistor becomes inoperative and is sometimes destroyed. The audio circuit designer avoids this hazard by maintaining a reasonably low base-circuit resistance. Note that "snowballing" occurs in the same manner in the case of a PNP transistor, except that the roles of electrons and holes are interchanged.

For convenience, graphs of the variation of collector current versus transistor temperature are reprinted in Fig. 4-5. The exponential spiral in Fig. 4-5a shows the general trend, and the family of curves depicted in Fig. 4-5b illustrates typical quantitative relations. Each curve is plotted with a fixed colector–base voltage ($V_{CB}$) and with a fixed emitter–base voltage ($V_{EB}$). We perceive that if collector-current variation versus temperature were caused only by saturation-current action, then the collector-current variation at temperatures below 10°C ($V_{EB} = 200$ mV and $V_{EB} = 300$ mV) would not occur. In fact, the collector current does vary with temperature, even when the saturation current is near zero. This variation is caused by the decrease in emitter–base junction resistance as the temperature increases. That is, the emitter–base junction resistance has a negative temperature coefficient of resistance.

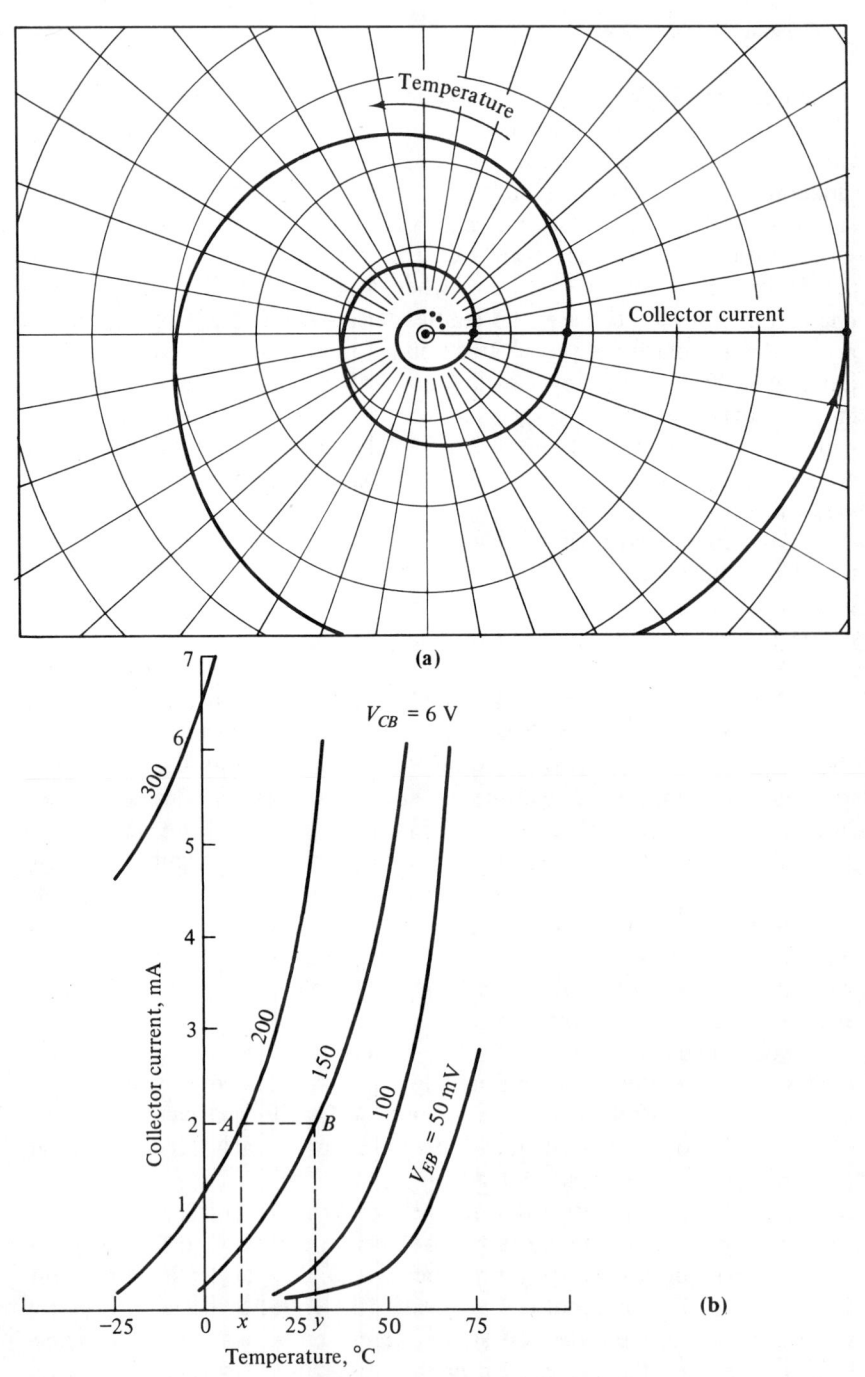

**Figure 4-5.** Saturation current increases exponentially with temperature: (a) characteristic increase; (b) plots of collector current versus temperature for various bias voltages.

## 4-2 Bias Stabilizing Circuits

A practical method of reducing the effect of the foregoing negative temperature coefficient of resistance is for the audio circuit designer to connect a high value of resistance in series with the emitter lead of the transistor. In effect, this series resistor causes the variation of the emitter–base junction resistance to be a small percentage of the total resistance in the emitter circuit. In other words, this external resistance swamps (overcomes) the emitter–base junction resistance; therefore, it is often termed a *swamping* resistor. Another practical method of reducing the effect of this negative temperature coefficient of resistance is also available to the audio circuit designer. He can utilize some means of reducing the emitter–base forward bias as the temperature increases. With reference to Fig. 4-5b, to maintain the collector current at 2 mA while the transistor temperature varies from 10°C (at $x$) to 30°C (at $y$), the forward bias must be reduced from 200 mV (at $A$) to 150 mV (at $B$). The temperature difference is 20°C (30 − 10); the voltage difference (200 mV − 150 mV) is 50 mV. In turn, the variation in forward bias per degree centigrade is calculated as follows:

$$\frac{\text{Difference in forward bias}}{\text{Difference in temperature}} = \frac{50 \text{ mV}}{20°C} = 2.5 \text{ mV}/°C$$

This equation indicates that the collector current will not vary with changes in emitter–base junction resistance if the forward bias is reduced 2.5 mV/°C for increasing temperature, or if it is increased 2.5 mV/°C for decreasing temperature.

## 4-2 Bias Stabilizing Circuits

Bias stabilizing circuits may utilize a variety of circuit components and devices from which the audio circuit designer may make a judicious choice (each will be discussed subsequently):

1. Common resistors may be employed. The effectiveness of external resistors in bias stabilization of transistor amplifiers is indicated by an analysis of the general bias circuit, as explained below. It will be seen that this analysis results in mathematical expressions for the current and voltage stability factors ($S_I$ and $S_V$) for each of three standard amplifier configurations.
2. Thermistors.
3. Junction diodes.
4. Transistors.
5. Breakdown diodes.

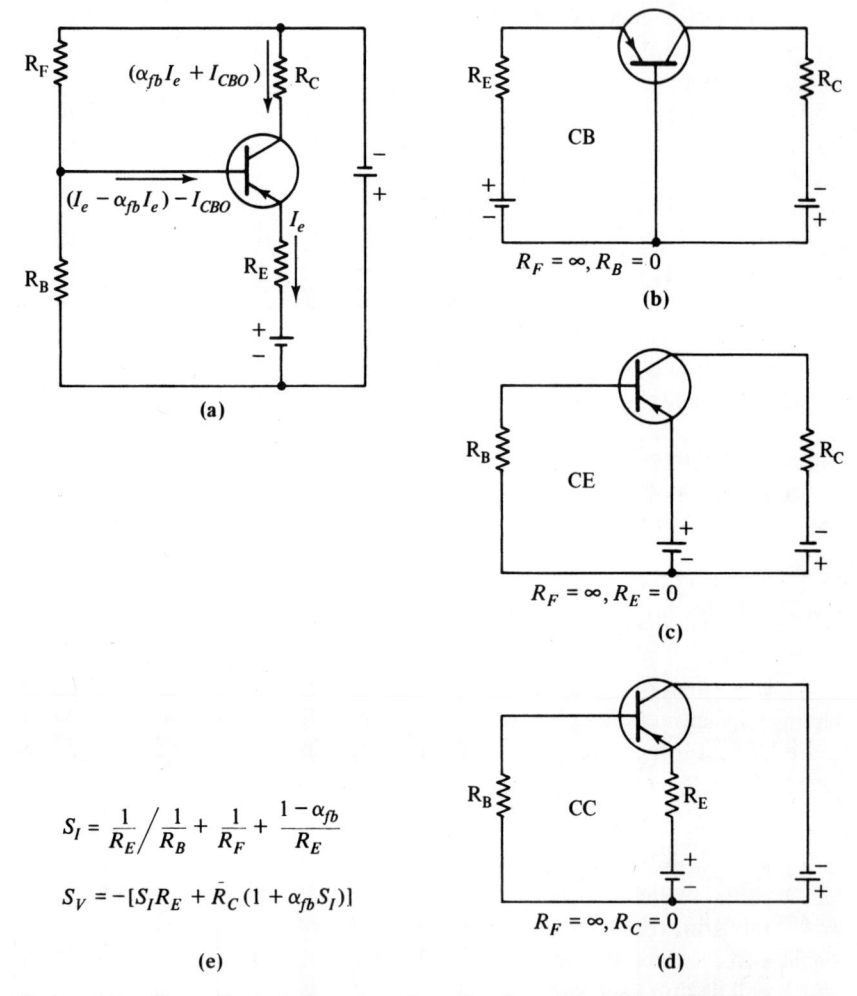

$$S_I = \frac{1}{R_E} \bigg/ \frac{1}{R_B} + \frac{1}{R_F} + \frac{1-\alpha_{fb}}{R_E}$$

$$S_V = -[S_I R_E + R_C(1 + \alpha_{fb} S_I)]$$

(e)

**Figure 4-6.** Generalized skeleton transistor circuit and derived configurations, with current and voltage stability factors.

## Resistor Stabilizing Circuits

It is helpful to consider the generalized skeleton transistor circuit shown in Fig. 4-6. Common-emitter, common-base, and common-collector (emitter-follower) configurations were noted previously. These are special cases of the generalized skeleton circuit. Depending upon the points of input and output of the signal into the generalized configuration and the values assigned to the resistors, the three basic modes of operation (CE, CB, and CC) can be derived.

## 4-2 Bias Stabilizing Circuits

1. *CB amplifier.* The CB amplifier in Fig. 4-6b is derived from the skeleton configuration by opening resistor $R_F$ ($R_F = \infty$), and by short-circuiting resistor $R_B$ ($R_B = 0$). The input signal is introduced into the emitter–base circuit, and it is extracted from the collector–base circuit.
2. *CE amplifier.* The CE amplifier in Fig. 4-6c is derived from the skeleton configuration by opening resistor $R_F$ and by short-circuiting resistor $R_E$. In this circuit, the signal is introduced by the base–emitter branch and it is extracted from the collector branch.
3. *CC amplifier.* The CC amplifier in Fig. 4-6d is derived from the skeleton configuration by opening resistor $R_F$ and by short-circuiting resistor $R_C$. The collector is common to the input and the output circuits. (Circuits in which $R_F$ is not open are discussed later.)

### Current Stability Factor

With reference to Fig. 4-6e, consider the current stability factor.

1. The ratio of a change in emitter current ($\triangle I_E$) to a change in saturation current ($\triangle I_{CBO}$) is a measure of the bias stability of a transistor. This ratio indicates the effect of a change in saturation current on the emitter current, and it is called the current stability factor $S_I$). The current stability factor is expressed by the equation

$$S_I = \frac{\triangle I_E}{\triangle I_{CBO}}$$

Inasmuch as the current stability factor is a ratio of two current values, it is expressed as a pure number. Under ideal conditions, the current stability factor would be equal to zero. In other words, the emitter current would not be affected by a change in saturation current.

2. By using the resistors and the indicated expressions for current flow in Fig. 4-6a, current $I_E$ can be expressed in terms of the saturation current $I_{CBO}$ and the resistive values employed in the circuit. Analysis of such an equation to indicate how current $I_E$ varies with saturation current $I_{CBO}$ results in the equation for the current stability factor ($S_I$) noted in Fig. 4-6e. This current stability factor refines the effectiveness of the external circuit to minimize emitter-current variation in a transistor.

3. To derive the CB amplifier configuration from the skeleton circuit in Fig. 4-6a, resistor $R_F$ is opened ($= \infty$) and resistor $R_B$ is short-circuited. To obtain the stability factor for this CB circuit, infinity is substituted for $R_F$ and zero is substituted for $R_B$ in the general equation for $S_I$. Observe that the reciprocal of infinity is zero, and that the reciprocal of zero is infinity. Accordingly, $S_I$ becomes zero. This is the ideal condition inasmuch as it minimizes the accumulation of minority carriers in the base region of the transistor, and permits the use of an emitter swamping resistor.

4. Direct substitution of the values of resistor $R_F$ ($= \infty$) and for resistor $R_E$ ($= 0$) in the general equation for $S_I$ with respect to the CE configuration depicted in Fig. 4-6c yields an indeterminate answer. That is, we obtain infinity divided by infinity. On the other hand, if we multiply the numerator and the denominator by $R_E$ before substitution, the value of the current stability factor is given:

$$S_I = \frac{1}{1 - \alpha_{fb}}$$

5. With reference to Fig. 4-6d, substitution of $R_F = \infty$ yields an approximate equation for the current stability factor of the CC configuration:

$$S_I = \frac{R_B}{R_E}$$

Otherwise stated, the value of $S_I$ in the CC arrangement depends on the ratio of base resistance to emitter resistance. This is another example of the basic principle that the higher the base resistance, the poorer is the current stability factor, and the higher the emitter resistance, the better is the current stability factor.

### Collector Current Stability Versus
### Base Resistance and Emitter Resistance

Variation of collector current with temperature for different values of emitter resistance and of base resistance is exemplified in Fig. 4-7. Observe that the worst-case condition occurs when the emitter and base resistances are both zero. A slight improvement is provided by a base resistance of 40 k$\Omega$. On the other hand, the optimum condition is provided by a substan-

## 4-2 Bias Stabilizing Circuits

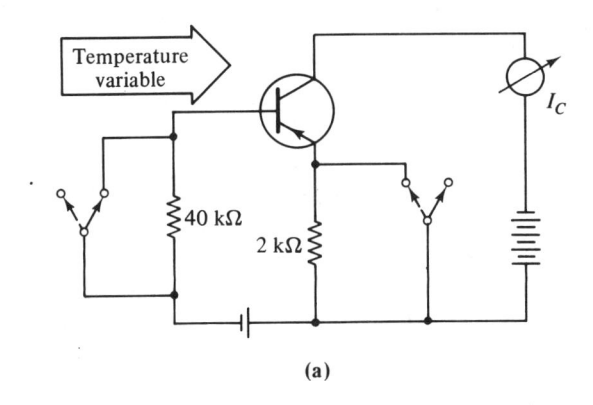

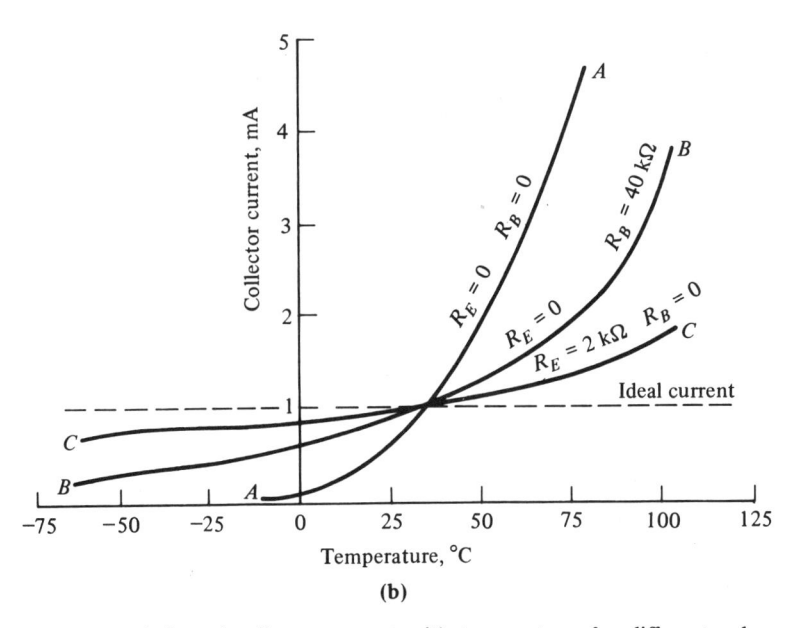

**Figure 4-7.** Variation of collector current with temperature, for different values of emitter resistance and base resistance: (a) configuration; (b) collector-current curves.

tial value of emitter resistance, with the base resistance equal to zero. A practical method of obtaining good current stability in the CE configuration is to use near-zero base resistance and an emitter swamping resistor, as depicted in Fig. 4-8a. Emitter resistor $R_E$ functions as a swamping resistor; the secondary winding of T1 introduces very little resistance into

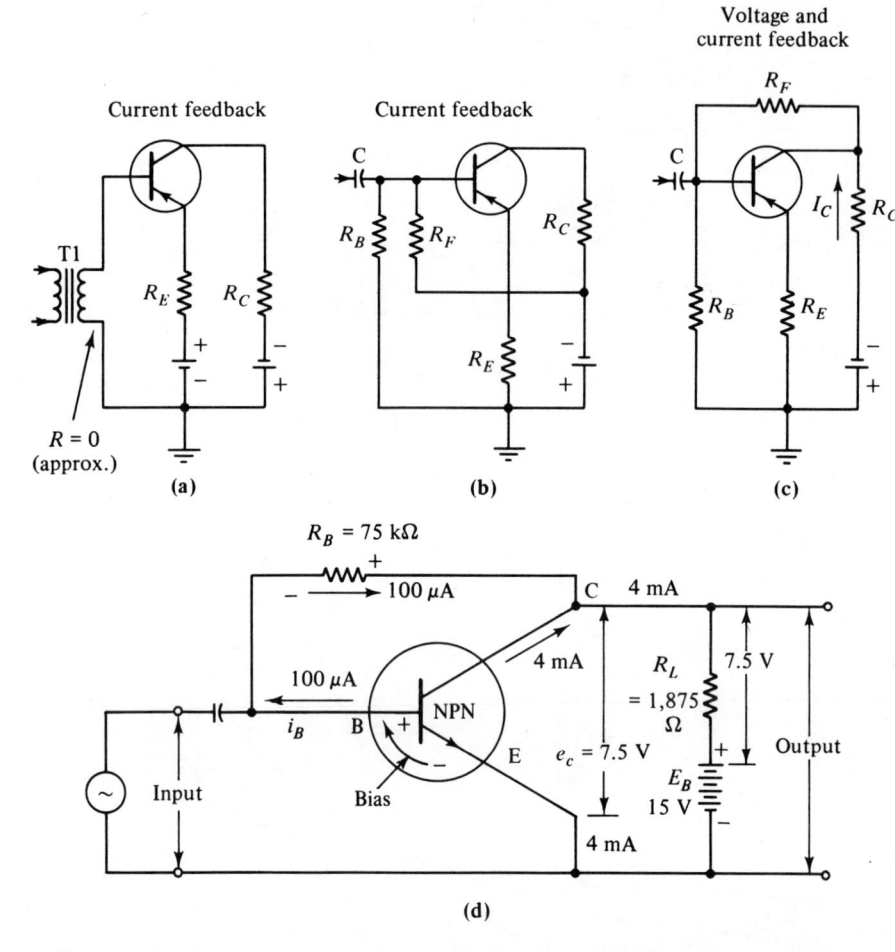

**Figure 4-8.** Common-emitter circuits that employ transformer input, fixed base voltage, and negative feedback: (a) emitter swamping resistor and negligible base resistance; (b) emitter swamping resistor with voltage-divider bias network; (c) emitter swamping resistor with voltage feedback; (d) example of voltage feedback without emitter swamping resistor.

the base circuit. The collector stability curve for this configuration is similar to that of curve *CC* in Fig. 4-7b. This circuit has a current stability factor of zero.

Refer to Fig. 4-8b. This arrangement is similar to that shown in Fig. 4-8a, except that a single battery is utilized, and RC coupling is employed instead of transformer coupling. A fixed emitter–base bias voltage is obtained by means of the voltage-divider network $R_F$–$R_B$. The base–emitter

## 4-2 Bias Stabilizing Circuits

bias voltage comprises the net voltage drop across $R_B$ and $R_E$. Substitution in the general current-stability equation of Fig. 4-6e yields the relation

$$S_I = \frac{R_B R_F}{R_B + R_F} \bigg/ R_E$$

Otherwise stated, the current-stability factor for the configuration in Fig. 4-8b is equal to the ratio of the net parallel-resistance value of $R_B$ and $R_F$ (base ground-return resistance) to the emitter resistance. Again, this result substantiates the basic principle that the lower the base-ground resistance and the higher the emitter resistance, the better is the current stability. There is a tradeoff in the lower value of $R_B$ that can be chosen, because the coupling capacitor C would have to be objectionably large. Even if an unusually large value of coupling capacitance were employed, an excessively low value of $R_B$ would result in heavy loading of the source (or the preceding stage).

We next consider a configuration that utilizes negative voltage feedback to improve the current stability factor, as exemplified in Fig. 4-8c. Note that if the collector current $I_C$ rises, the collector potential becomes less negative because of the larger dc voltage drop across $R_C$. Hence, less forward bias voltage (negative base to positive emitter is coupled through resistor $R_F$ to the base). Reduction of the forward bias voltage results in reduced collector current. A quantitative example of stage design with voltage feedback, but without an emitter swamping resistor, is shown in Fig. 4-8d. The dc input resistance to the base is 75 k$\Omega$, in accordance with Ohm's law. Since an emitter swamping resistor is not used, the bias stability of this configuration is not as good as that of the configuration shown in Fig. 4-8c.

A low-level audio-amplifier circuit that uses an emitter swamping resistor with voltage feedback and with a voltage-divider bias network is exemplified in Fig. 4-9. The quiescent current of this configuration is 1 mA; the beta value of the transistor is at its maximum with this value of quiescent current. The stability factor is high, owing to the shunt degeneration provided by R3 and to the series degeneration provided by R2. This circuit operates satisfactorily over a temperature range from $-60°C$ to $+75°C$. To obtain maximum dynamic range, the voltage drop across R4 is designed to be approximately one half of the supply voltage. Resistor R2 is bypassed to avoid ac signal degeneration. The degeneration introduced by R3 linearizes the transfer characteristic, reduces the current gain, reduces the input and output impedances of the stage, and extends the frequency response. A greater amount of negative feedback results when $R_L$ is high or when $R_S$ is high. The specifications for this amplifier

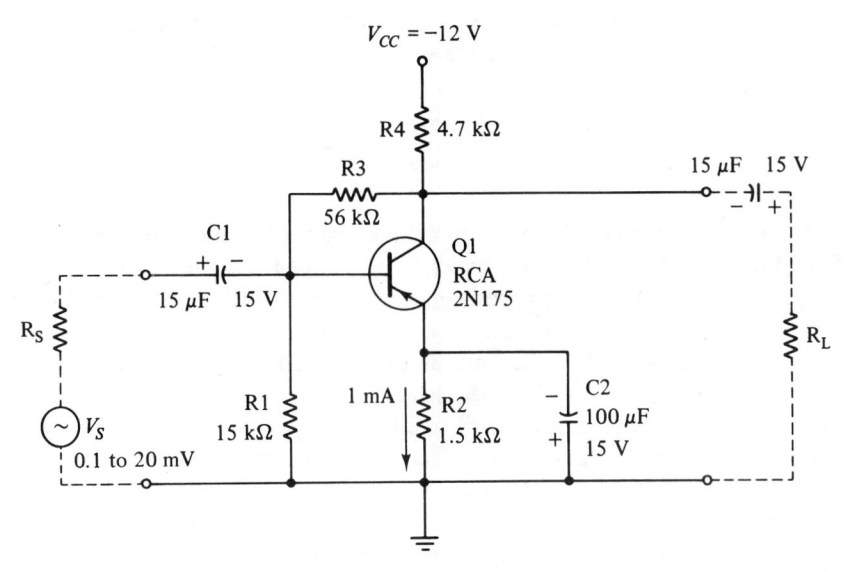

**Figure 4-9.** Low-level audio amplifier that uses an emitter swamping resistor with voltage feedback and with a voltage-divider network.

are listed in Table 4-1. For minimum distortion, a 5-kΩ load and a 1-mV signal input should be utilized.

**Table 4-1. Specifications for the Amplifier Design of Figure 4-9**

| | | |
|---|---|---|
| *Frequency response:* | A. | Current gain vs. frequency (10 to 40,000 Hz at 3 dB down) |
| | B. | Voltage gain vs. frequency (55 to 10,000 Hz at 3 dB down) |
| *Minimum input level:* | 0.1 mVac | |
| *Maximum input level:* | 20 mVac | |
| *Distortion:* | Maximum % total harmonic distortion | |

| | $R_S$ 5,000 Ω | $R_S$ 1,000 Ω |
|---|---|---|
| *Input voltage:* | | |
| A.    1 mVac | 0.7% | 1.0% |
| B.   10 mVac | 3.9% | 6.2% |
| C.   20 mVac | 8.3% | 13.0% |
| *Gain at 30°C* | *Current gain* | *Voltage gain* |
| A.   $R_L = 2,000$ Ω | 13 | 36 |
| B.   $R_L = 500$ Ω | 41 | 16 |

## 4-3 Thermistor Stabilization

**Table 4-1.** (*Continued*)

*Characteristic impedance at 30°C*

A. Output impedance = 1,620 Ω when $R_S$ = 1,000 Ω
                              980 Ω when $R_S$ = 5,000 Ω

B. Input impedance = 1,620 Ω when $R_L$ = 500 Ω
                               900 Ω when $R_L$ = 2,200 Ω

*Power requirements:* −12 V ± 10% at 1 mAdc 12 mW

*Temperature range:* −60 to +75°C

### 4-3 Thermistor Stabilization

To prevent the emitter current of a transistor from increasing with a rise in ambient temperature, an external circuit may be used that includes a temperature-sensitive device or devices. A thermistor is an example of such a device. Its resistance value decreases as the ambient temperature increases; that is, it has a negative temperature coefficient of resistance. We will consider a circuit arrangement that uses a thermistor to provide emitter voltage control. With reference to Fig. 4-10, an amplifier configuration is exemplified that includes a thermistor to vary the emitter voltage with varying temperature to minimize resultant changes in emitter current. The thermistor is selected by the audio circuit designer to provide an optimum control characteristic.

This circuit comprises two voltage dividers; the first consists of R4 and R1, and the second consists of R2 and thermistor RT1. The first voltage divider permits application of a portion of the battery voltage ($V_C$) to the base terminal and ground (common return). The base terminal voltage is developed across R1 and is in the forward bias polarity. The second voltage divider applies a portion of battery voltage $V_C$ to the emitter terminal. This emitter terminal voltage is developed across R2 and is in the reverse bias polarity. Since the forward bias voltage applied to the base terminal is larger than the reverse bias applied to the emitter terminal, the resultant base–emitter bias is always in the forward direction.

With an increase in temperature, the collector current in Fig. 4-10 would increase if an unstabilized circuit were employed. However, this circuit functions to reduce the forward bias on the transistor as the temperature increases. This reduction is accomplished by action of the voltage divider comprising R2 and thermistor RT1. When the temperature increases, the resistance of RT1 decreases, and more current flows through the voltage divider. This increased current flow raises the negative potential

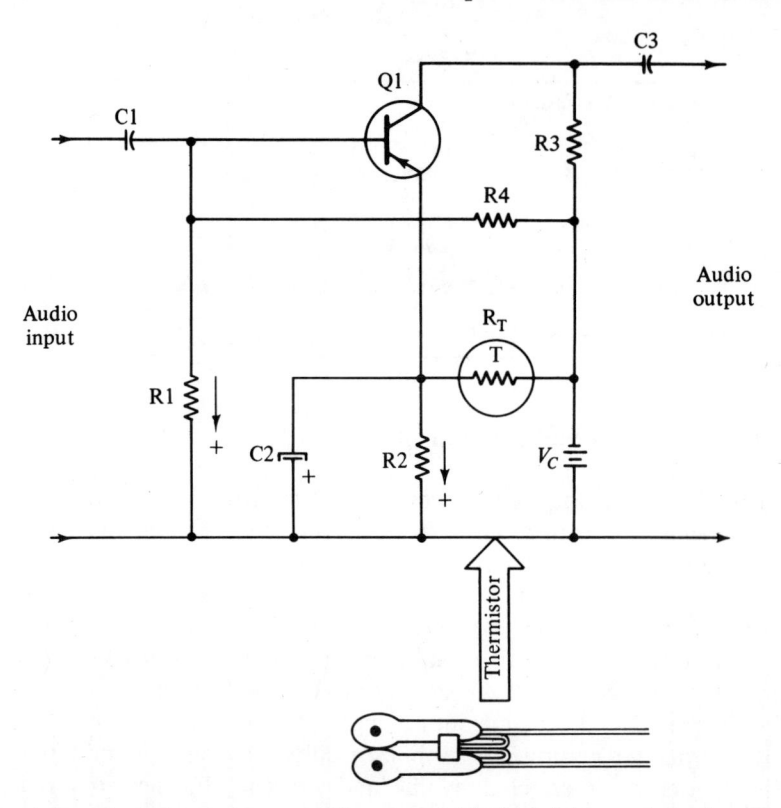

**Figure 4-10.** Typical transistor amplifier stage with thermistor control of emitter bias voltage.

at the emitter connection of R2. Thereby, the reverse bias voltage applied to the emitter increases, and the net emitter–base forward bias decreases. Accordingly, the collector current is reduced. On the other hand, if the temperature decreases, reverse circuit action ensues to prevent decrease of collector current. The audio circuit designer must analyze worst-case conditions on the basis of both transistor and thermistor tolerances, as well as resistor tolerances.

Capacitor C1 in Fig. 4-10 serves to block the dc voltage from the preceding stage, and it couples the ac signal voltage into the base–emitter circuit of the transistor. Capacitor C2 bypasses the ac signal around R2. Resistor R3 is the collector load resistor that develops the output signal voltage. Capacitor C3 blocks the dc collector voltage from and couples the ac signal voltage to the next stage. It is good practice to mount the thermistor next to the transistor, so that both devices will be subjected to the same ambient temperature. Otherwise, the circuit designer may fail to realize the full control potential of the thermistor.

## 4-3 Thermistor Stabilization

### Base-Voltage Thermistor Control

Refer to Fig. 4-11. Here a thermistor is employed to vary the base voltage of the transistor in response to temperature change. Thereby, emitter current variation is minimized. This configuration includes a voltage divider comprising R1 and thermistor RT1. The voltage divider applies a portion of battery voltage $V_C$ to the base–emitter circuit of the transistor. Electron flow through the voltage divider occurs in the direction of the arrow. This current produces a voltage of the indicated polarity across thermistor RT1. A forward bias is applied to the transistor. In the event of temperature increase, the emitter current would tend to rise in an unstabilized circuit. However, in this configuration, the resistance of thermistor RT1 decreases as the temperature rises, thereby causing increased current flow through the voltage divider. This increased current flow causes a larger portion of the battery voltage $V_C$ to drop across R1. Consequently, the available voltage for forward bias, of the drop across RT1, decreases and the emitter current is thereby reduced.

This is an example of transformer-coupled circuitry in which the base resistance is very small. Although the reactance of capacitor C1 is in series with the base lead, its reactance value will be very low if the circuit designer chooses a sufficiently large value of capacitance. Capacitor C1 functions to bypass the ac signal around thermistor RT1. The primary winding of transformer T2 operates as the collector load; it develops the output signal voltage, which in turn is coupled to the secondary winding

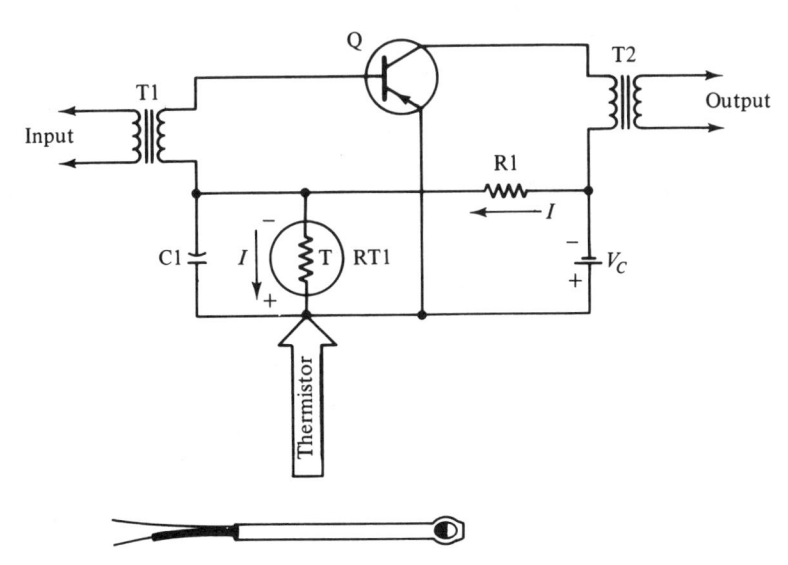

**Figure 4-11.** Amplifier-stage configuration with thermistor control of base bias voltage.

on T2. The audio circuit designer must make his worst-case analysis for this bias stabilization arrangement from the viewpoint of tolerances on both Q and RT1, in addition to resistor tolerances.

## Thermistor Capabilities and Limitations

Although thermistor circuits have considerable capability in control of collector current versus temperature, they also have certain limitations. Thermistor capability in the configurations of Figs. 4-10 and 4-11 is graphed in Fig. 4-12. An unstabilized amplifier circuit is shown in Fig.

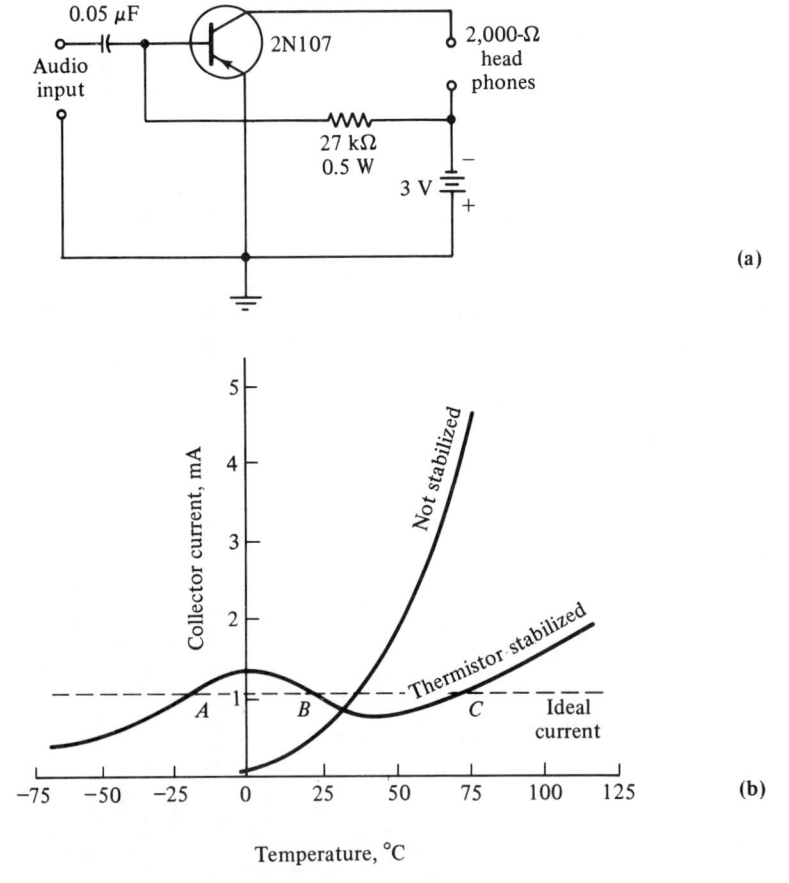

**Figure 4-12.** Variation of collector current versus temperature for nonstabilized and for temperature-stabilized amplifier stages: (a) unstabilized transistor amplifier stage; (b) curves of collector current versus temperature for unstabilized and for stabilized configurations.

4-12a, and its variation of collector current versus temperature is indicated by the "not stabilized" curve in Fig. 4-12b. On the other hand, the "thermistor stabilized" curve in Fig. 4-12b shows the control of collector current that is provided by the arrangements in Figs. 4-10 and 4-11. The improvement in bias stability as a result of thermistor control is apparent. Note, however, that thermistor stabilization can realize ideal stabilization at only three temperatures—at points *A, B,* and *C* in Fig. 4-12b. Otherwise stated, thermistor resistance variation does not precisely track transistor emitter–base junction resistance variation. Consequently, the audio circuit designer may prefer another type of bias stabilization circuit that provides improved tracking.

## 4-4 Diode Bias Stabilization Circuits

Variation of collector current in a transistor versus temperature is caused by a corresponding variation in emitter–base junction resistance and the saturation (reverse bias) current. Variation with temperature in the resistance and the reverse bias current of a PN junction occurs whether the junction is in a transistor or in a junction diode. The voltage–current characteristic is rather similar for both diode and transistor junctions. Therefore, a junction diode is a useful device for the audio circuit designer to utilize in bias stabilization circuits. Another advantage of a diode in this application is that the designer can choose a type of diode that has the same substance as used in the controlled transistor, either germanium or silicon. The temperature coefficient of resistance for the diode and for the resistor will then be virtually the same. In turn, better tracking of the bias stabilization circuit can be obtained.

With reference to Fig. 4-13, the lattice structure of a germanium diode is represented. The junction is flanked by N-type and P-type material. When a negative potential is applied to the P-type material and a positive potential to the N-type material, no current flows across the junction in theory. However, in practice, a small amount of reverse current (saturation or leakage current) does flow, and this current increases exponentially with rising temperature. Junction diodes have a negative temperature coefficient of resistance, whether they are forward or reverse biased. Consider next how a single forward-biased diode can be utilized to obtain bias stabilization in a transistor amplifier stage.

With reference to Fig. 4-13b, the circuit employs a forward-biased junction diode CR1 as a temperature-sensitive device for compensation of emitter–base junction resistance variations in Q. Current ($I$) through the voltage divider flows in the indicated direction, and it develops a

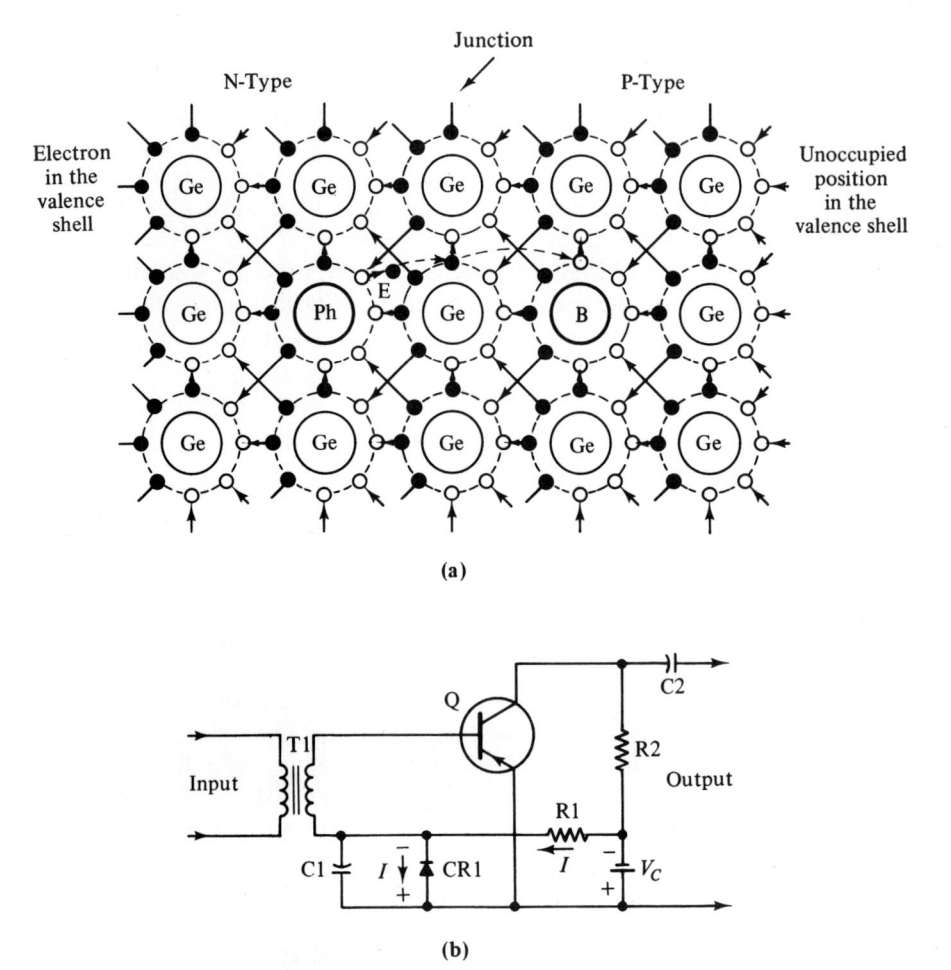

**Figure 4-13.** Transistor amplifier stage with a single forward-biased junction diode for compensation of emitter–base junction resistance with temperature: (a) lattice structure of germanium diode; (b) amplifier configuration.

voltage drop across CR1 of the indicated polarity. This voltage drop provides a forward bias for Q. With a temperature increase, the collector current would tend to increase in an unstabilized circuit. On the other hand, an increase in temperature decreases the resistance of CR1, and in turn causes more current to flow through the voltage divider. Consequently, there is an increased voltage drop across R1. The voltage drop across CR1 is correspondingly decreased, with the result that the forward bias voltage is reduced, and the collector current through Q is decreased.

## 4-4 Diode Bias Stabilization Circuits

The input ac signal is coupled into the transistor by transformer T1. Capacitor C1 bypasses the ac signal around CR1. Output signal voltage is developed across the collector load resistor R2. Capacitor C2 blocks dc voltage from and couples the ac signal voltage into the following stage. Refer to Fig. 4-14a. Electrons flow from the emitter region into the base region, and diffuse across the base substance into the collector region. The flow of electrons from the emitter substance into the base substance is

(a)

(b)

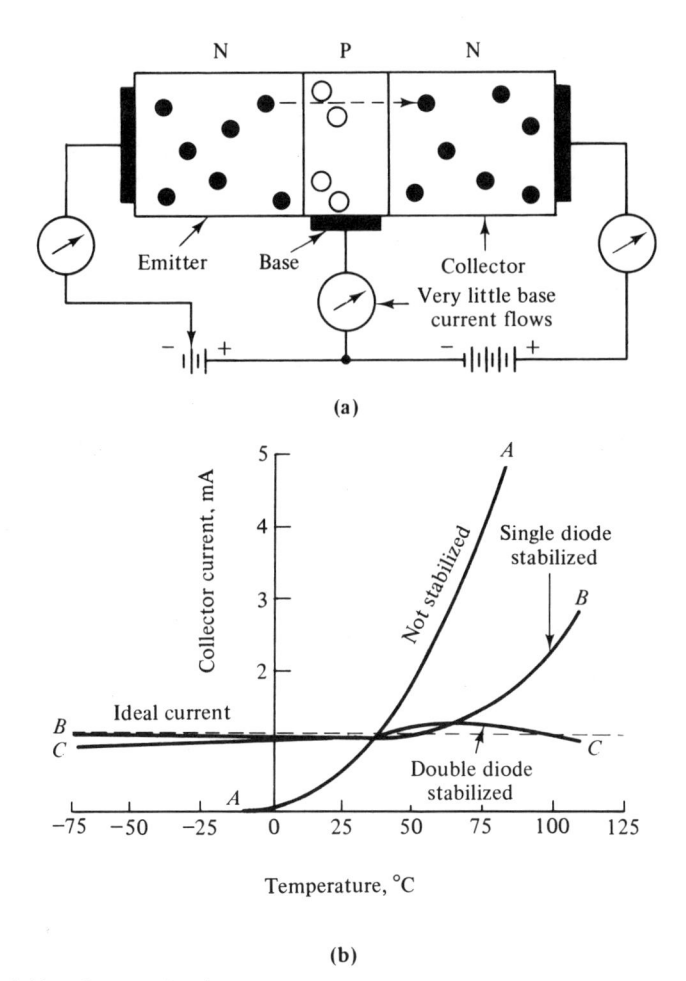

**Figure 4-14.** Graphs showing the variation of collector current versus temperature for nonstabilized, single-diode stabilized, and double-diode stabilized transistor amplifier circuits: (a) electrons flow from the emitter region into the base region, and diffuse through the base region into the collector region; (b) graphs of collector current versus temperature for stabilized and unstabilized bias circuits.

depicted in Fig. 4-13a. Next, the effectiveness of the diode bias stabilization circuits is indicated by curve *BB* in Fig. 4-14b. Observe that curve *BB* exhibits a marked improvement in collector-current stability for temperatures below 50°C; in other words, the variation of the diode resistance with temperature tracks with variations in emitter–base junction resistance more precisely than does a thermistor bias stabilization circuit. Curve *CC* is considered next.

The sharp increase in collector current (curve *BB*) at temperatures above 50°C indicates that the junction diode CR1 does not compensate for the increase in saturation current ($I_{CBO}$). This condition is to be expected, inasmuch as the saturation current (collector–base reverse bias current) flows out of the base, through the primary of T1, through diode CR1, through battery $V_C$, and back to the collector. Since the saturation current value is a small percentage of the total current flow through diode CR1 (forward biased and with a low junction resistance), it causes no appreciable voltage drop across diode CR1. Therefore, the audio circuit designer must employ a second junction diode (reverse biased) to compensate for variation in saturation current, as explained next.

## Double-Diode Stabilization

Two semiconductor diodes are employed as temperature-sensitive devices in the amplifier configuration of Fig. 4-15b. Forward and reverse current flows are depicted in Fig. 4-15a. One junction diode compensates for temperature variation in saturation current. The other diode compensates for temperature variation in emitter–base junction resistance. Resistor R3 and junction diode CR2 (reverse biased) are included. Resistor R1 and diode CR1 (forward biased) compensate for change in emitter–base junction resistance at temperatures below 50°C. Reverse-biased diode CR2 can be considered an open circuit at low temperatures. At room temperature, the CR2 reverse-bias current $I_s$ flows through CR2 in the direction indicated. Diode CR2 is selected by the audio circuit designer so that its reverse-bias current $I_s$ is larger than the reverse-bias current $I_{CBO}$ of the transistor. Diode reverse-bias current $I_s$ consists of transistor reverse-bias current $I_{CBO}$ and a component ($I_1$) drawn from the battery. The voltage polarity developed by $I_1$ across R3 is indicated. Note that the emitter–base bias voltage is the sum of the opposing voltages across R3 and CR1, assuming that the secondary winding of T1 has negligible resistance.

As the temperature increases, currents $I_{CBO}$, $I_s$, and $I_1$ increase. The resultant reverse-bias voltage dropped across R3 by current $I_1$ also increases. Thus the total forward bias (voltage across CR1 and R3)

## 4-4 Diode Bias Stabilization Circuits

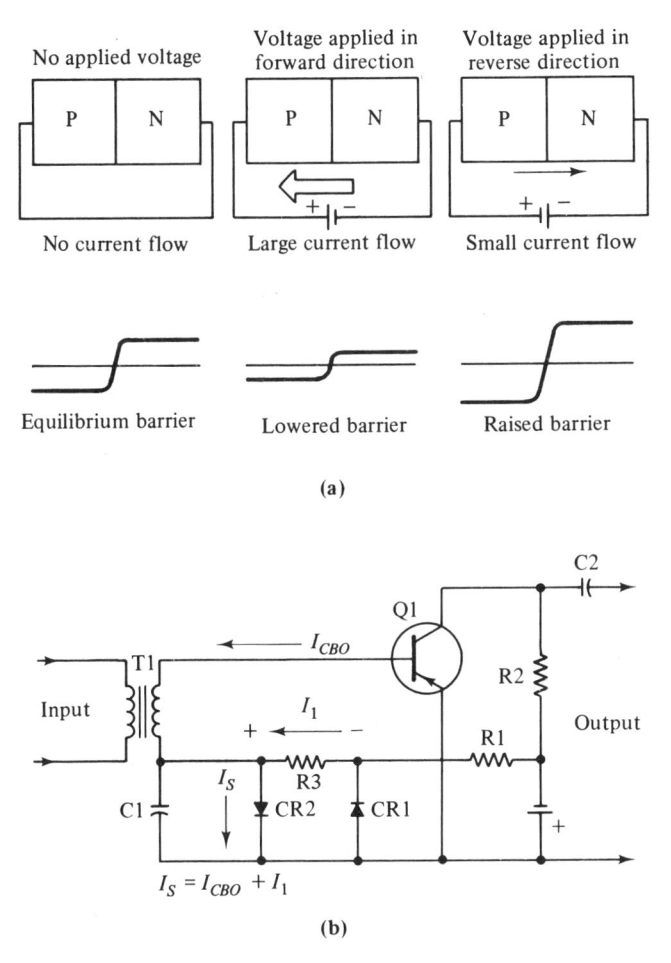

**Figure 4-15.** Amplifier with double-diode bias stabilization against temperature variation: (a) diode current flow; (b) bias circuitry.

decreases with increasing temperature to stabilize the collector-current flow. In a worst-case analysis of an amplifier configuration that includes this bias stabilization method, the audio circuit designer must take into account the tolerances on both of the diodes, on the transistor, and on the resistors.

### Bias Stabilization by a Single Reverse-Bias Diode

A transistor amplifier configuration that utilizes a single reverse-biased diode for compensation of collector current changes resulting from temperature variation is depicted in Fig. 4-16b. It will be seen from the rela-

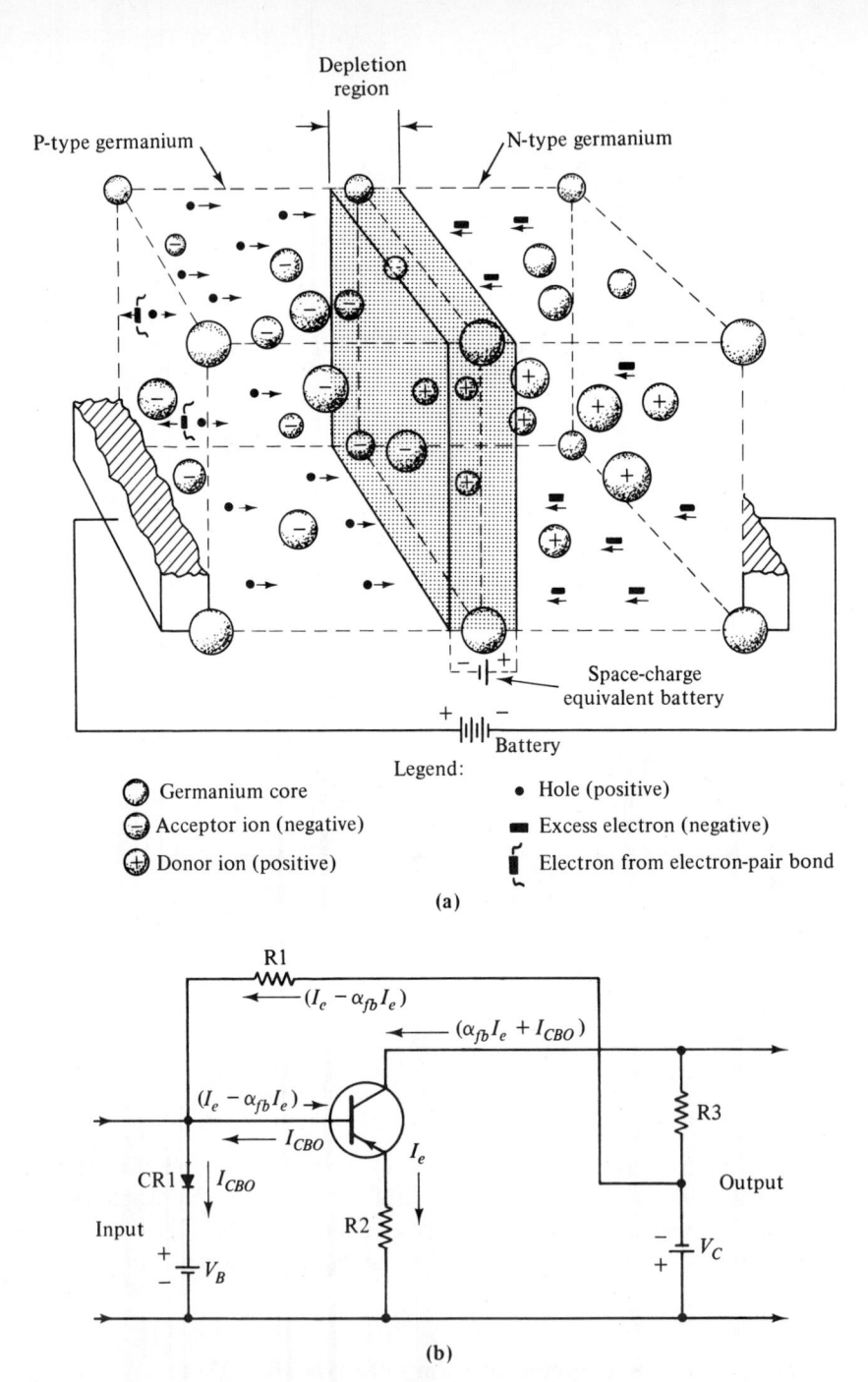

**Figure 4-16.** Transistor amplifier configuration that utilizes a single reverse-biased diode for temperature stabilization in saturation current: (a) holes are majority carriers and electrons are minority carriers in the P substance; electrons are majority carriers and holes are minority carriers in the N substance; (b) configuration.

tions in Fig. 4-16a that holes are majority carriers and electrons are minority carriers in P substance; conversely, electrons are majority carriers and holes are minority carriers in N substance. The exemplified configuration provides two separate paths for the two components of the base current. Thus the base–emitter current ($I_e - \alpha_{fb}$) flows through the base region to the emitter, through R2, battery $V_C$, and R1 to the base lead. Also, the saturation current ($I_{CBO}$) flows from the base lead through CR1, battery $V_B$, and the collector region to the base region. The audio circuit designer selects the diode that is utilized so that its saturation (reverse bias) current equals that of the transistor over a wide temperature range.

As the temperature increases, current $I_{CBO}$ also increases. On the other hand, the saturation current of CR1 increases by an equal amount, so that there is no accumulation of $I_{CBO}$ charge carriers in the base region. Such an accumulation would cause an increase in emitter current. Diode CR1 functions as a gate, and it opens wider to accommodate the increase in $I_{CBO}$ with rising temperature. This configuration is employed by the circuit designer if RC coupling is utilized between it and the previous stage. That is, the reverse-biased diode provides high resistance for the input circuit. Resistor R2 swamps the emitter–base junction resistance, and thereby prevents a large increase in emitter current, particularly at low temperatures. Resistor R3 is the collector load resistor; it drops the ac output signal voltage.

## 4-5 Transistor Bias Stabilization Circuits

It has been previously mentioned that some terminal voltages and currents of a temperature-stabilized transistor can be utilized to temperature stabilize another transistor (or several transistors). For example, a common emitter–base voltage may be employed for this purpose. Since the emitter–base junction of a transistor has a negative temperature coefficient of resistance similar to that of a PN junction diode, it is feasible to use the variations in emitter–base junction resistance in one transistor to control the emitter–base bias of another transistor under conditions of temperature change. A configuration that exploits this principle is exemplified in Fig. 4-17. Here, the emitter–base voltage from Q1 is also used to bias the emitter–base junction of Q2. If we assume zero dc resistance in the secondary winding of T2, the base of Q2 may be regarded as connected directly to the emitter of Q1. Again, if we assume zero dc resistance in the secondary of T1, the emitter of Q2 may be regarded as directly connected to the base of Q1. Battery $V_E$ supplies forward bias for Q1. Cross-connection of emitters and bases for the two transistors results in forward bias on Q2, since it is a PNP-type transistor.

**Legend:**

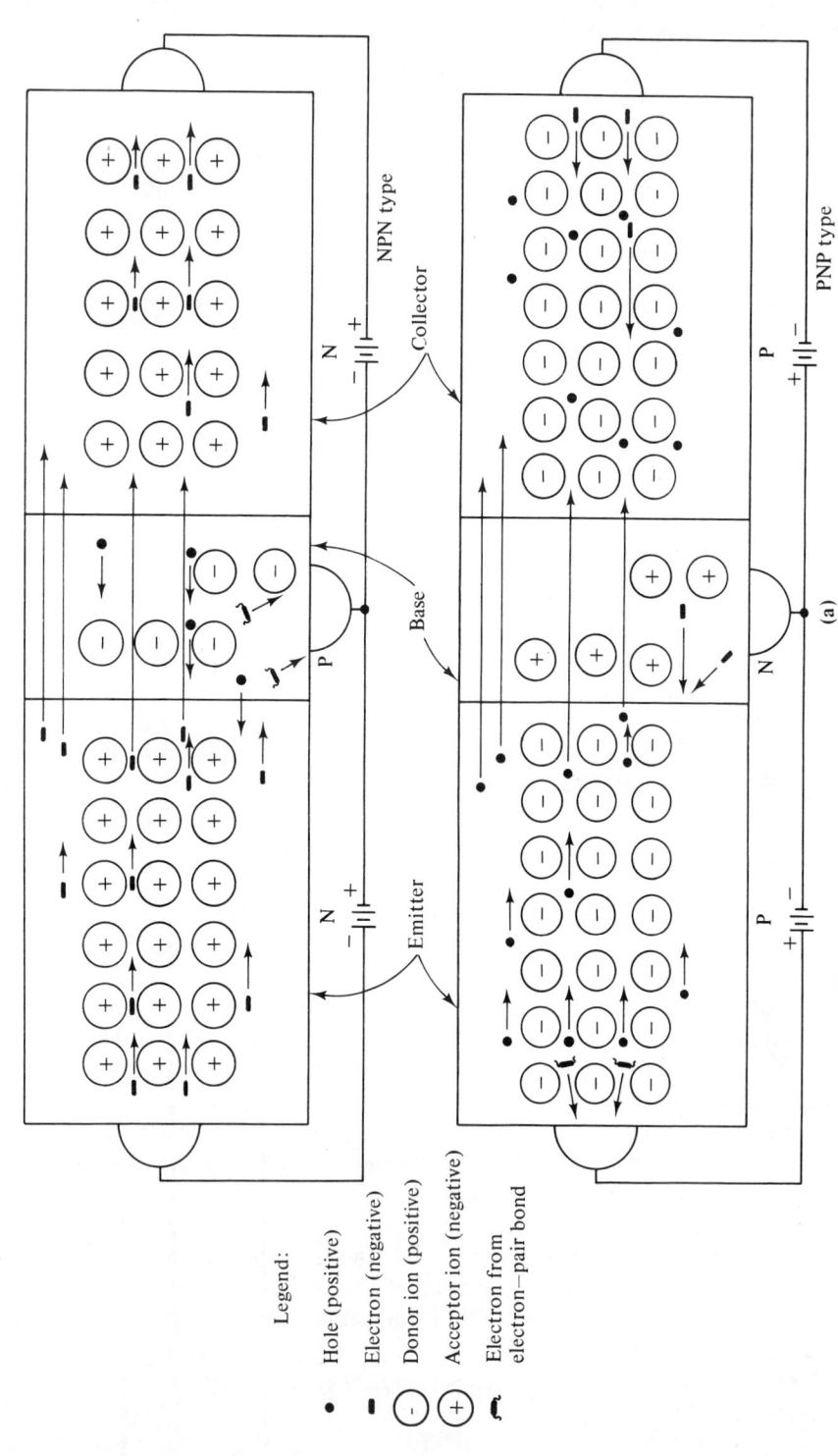

Hole (positive)
Electron (negative)
Donor ion (positive)
Acceptor ion (negative)
Electron from
electron–pair bond

**Figure 4-17.** An amplifier configuration that employs the emitter-base voltage of Q1 for bias stabilization of Q2: (a) charge-carrier relations in the application of forward bias between emitter and base, and reverse bias between base and collector.

## 4-5 Transistor Bias Stabilization Circuits

Two-stage amplifier

(b)

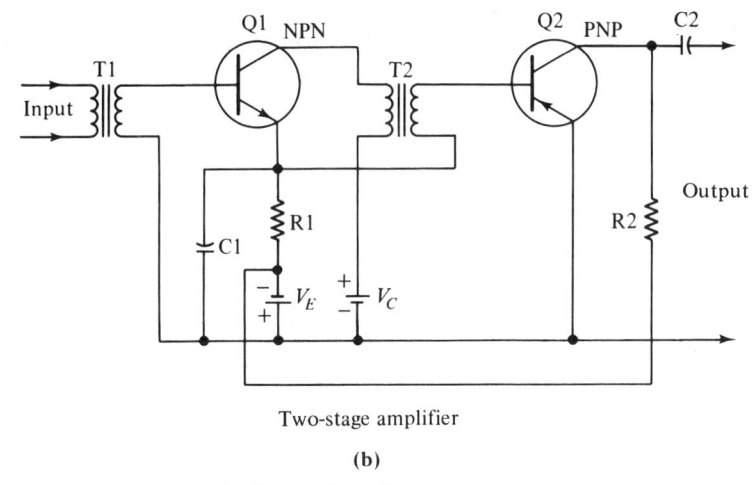

**Figure 4-17.** (*Continued*) (b) configuration.

Transistor Q1 is temperature stabilized by insertion of a high-valued resistor R1 (swamping resistor) in its emitter branch; stabilization is enhanced by low dc resistance in the base branch. Resistor R1 functions to maintain a relatively constant emitter current for Q1. As the temperature rises, the base–emitter-junction resistance of Q1 decreases. Since the current through the junction remains constant, the voltage drop across the junction decreases. This decrease represents a decrease in forward bias for Q2. Accordingly, the tendency of the collector current through Q2 to increase with rising temperature is compensated. If a curve of collector current through Q2 were plotted versus temperature, it would be closely similar to curve *CC* shown in Fig. 4-14b.

### Common Emitter–Collector Bias Stabilization Method

Two-stage amplifier configurations with and without bias stabilization are shown in Fig. 4-18. When identical stages are cascaded, they are termed iterated stages. Source and load-resistance values of 10 k$\Omega$ are suggested in Fig. 4-18a, because this value provides near-maximum flexibility in assignment of values for $R_A$ and $R_B$. In Fig. 4-18b, a method of temperature stabilizing the emitter–collector current of one transistor by using the stabilized emitter–collector current of another transistor (instead of the stabilized emitter–base voltage) is shown. A swamping resistor (R2) is utilized for stabilization of Q1's collector current. The audio circuit designer selects a suitable value for R1 (in the low range) in order to minimize accumulation of saturation-current charge carriers in the base region of the transistor. The stabilized dc collector current

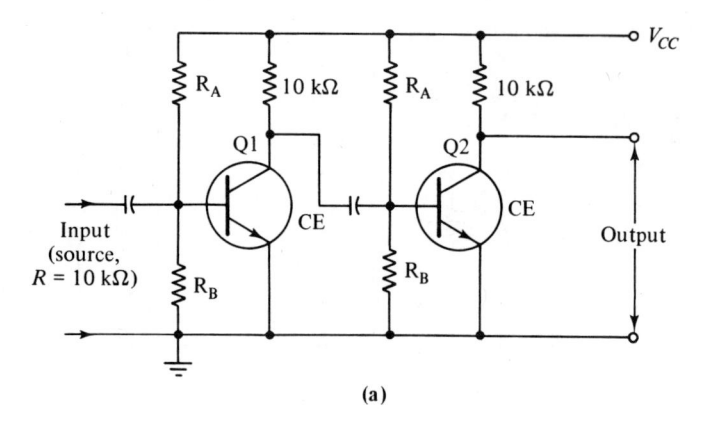

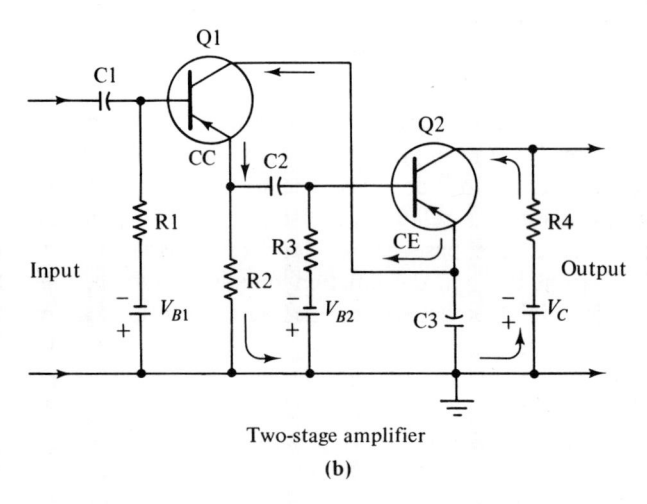

**Figure 4-18.** Two-stage amplifier configurations with and without bias stabilization: (a) iterated stages with no bias stabilization; (b) a two-stage configuration with employment of common emitter–collector current for temperature stabilization of both transistors.

of Q1 is caused to flow through the emitter–collector region of Q2 by connection of the collector of Q1 directly to the emitter of Q2. The current direction in this circuit is indicated by the arrows. Electrons that flow into the emitter lead of Q1 flow through R2, battery $V_C$, R4, collector–emitter of Q2, and into the collector of Q1.

An advantage of the foregoing method of stabilizing the collector current of Q2 is that an emitter swamping resistor is not required in Q2's emitter lead. Elimination of the emitter resistor is particularly desirable in power-amplifier stages that draw heavy emitter currents. Observe that

## 4-5 Transistor Bias Stabilization Circuits

Q1 is employed in a CC configuration, with Q2 in a CE configuration. Capacitor C3 holds the collector of Q1 and the emitter of Q2 at ac ground potential. Capacitor C1 functions as a dc blocking capacitor and couples the incoming ac signal to the base of Q1. Resistor R1 provides a dc return path for base current flow. Battery $V_{B1}$ provides base–emitter bias voltage. R2 functions as an emitter load resistor, develops the output signal voltage from Q1, and provides negative feedback. Capacitor C2 blocks the emitter dc voltage from and couples the ac signal to the base of Q2. Resistor R3 provides a dc return path for base current. Battery $V_{B2}$ supplies base bias voltage. Resistor R4 is the collector load resistor for Q2 and develops the output signal voltage. Battery $V_C$ supplies collector voltage for both transistors.

### Collector-Current Temperature Stabilization

With reference to Fig. 4-19, two stages of direct-coupled amplification are exemplified. A dc amplifier steps up a dc input voltage and very low frequency ac signals, in addition to the higher-frequency range that has been noted for RC-coupled amplifiers. Since no capacitors are utilized in this configuration, it has equal gain for a dc input voltage and for a midband ac signal voltage. This circuit is designed so that an increase in Q1's collector current caused by a temperature rise will reduce the forward bias on Q2. Its operation is as follows: transistor Q1 is connected in the CB arrangement; therefore, its stability factor is ideal. On the other hand, there will be some variation in Q1's collector current with rising temperature. Let us assume that a collector-current increment of $\Delta I_C$ occurs. The direction of this current increase is indicated by the arrow. A portion of this increment in collector current flows through R3. This portion is designated $\Delta I_{C1}$, and it develops a voltage across R3 with the indicated polarity.

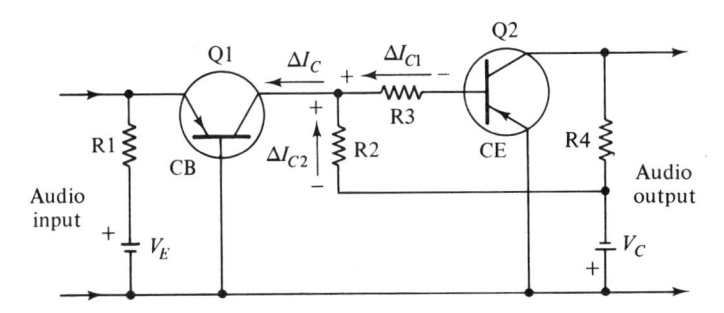

**Figure 4-19.** Two-stage direct-coupled amplifier configuration with collector-current temperature stabilization.

Another portion of the increment in collector current flows through R2. This portion is designated $\triangle I_{C2}$, and it develops a voltage across R2 with the indicated polarity. Observe that the voltage polarities indicated are only for an increment in collector current; they are not necessarily steady-state voltage polarities. The steady-state voltages are of no concern to the audio circuit designer in a bias stabilization analysis. Consider next the base–emitter bias circuit for Q2. Its base–emitter bias voltage is equal to the sum of the voltages dropped across R3 and R2 and the voltage of battery $V_C$. The voltage indicated across R3 aids the forward bias; conversely, the voltage indicated across R2 opposes the forward bias. The audio circuit designer selects values for R2 and R3 such that R2 drops a larger voltage; thereby, the resultant forward bias is decreased. This circuit action limits the tendency of Q2's collector current to increase with rising temperature.

R1 in Fig. 4-19 functions as an emitter swamping resistor. Battery $V_E$ supplies emitter bias voltage. Resistor R4 functions as a collector load resistor. The voltage stability factor for the stage is expressed by an equation that was noted earlier in this chapter; inspection will show that the voltage stability factor is directly proportional to the current stability factor (an equation that was also noted earlier). In turn, the circuit techniques that improve the current stability factor also improve the voltage stability factor. Otherwise stated, variations in collector voltage with changes in temperature are limited as a consequence of variations in saturation current with changes in temperature.

## 4-6 Temperature Compensation by Breakdown (Zener) Diodes

A reverse-biased diode junction has an $E/I$ characteristic as exemplified in Fig. 4-20. The diode maintains a comparatively constant reverse-current value versus voltage up to a critical point ($E_{\text{out}}$), after which current increase ensues rapidly, although the voltage drop across the diode remains essentially constant. This is a constant-voltage source, as provided by zener-diode action, and it has various useful applications in bias stabilization arrangements. The basic voltage-regulator circuit depicted in Fig. 4-20b maintains a constant load voltage, although the source voltage or the load current may vary over a considerable range. Typical breakdown voltages for commercial zener diodes are from 2 to 60 V.

### Zener-Diode Temperature Compensation

A reverse-biased diode junction has a negative temperature coefficient of resistance. Note that this characteristic is limited to a reverse-bias voltage

## 4-6 Temperature Compensation by Breakdown (Zener) Diodes    147

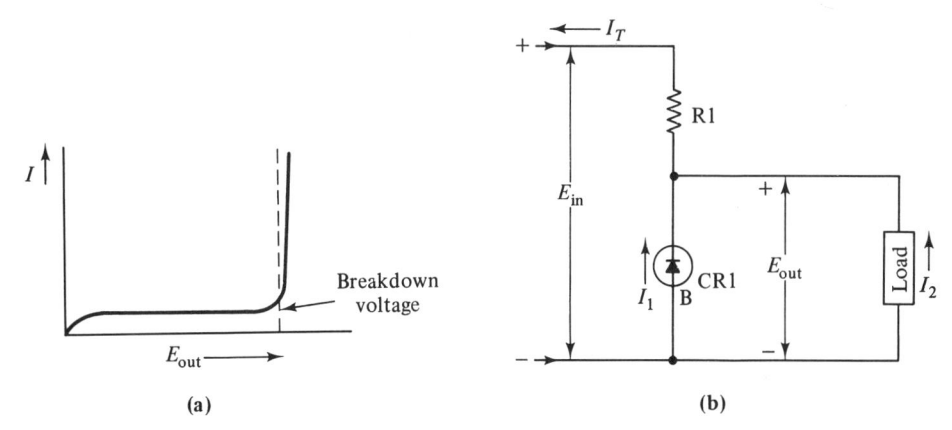

**Figure 4-20.** Breakdown (zener) diode voltage-current characteristic and voltage-regulator circuit.

that does not equal or exceed the breakdown voltage value. A zener diode has a positive temperature coefficient of resistance several times larger than the negative temperature coefficient of resistance for a forward-biased or a reverse-biased junction diode. Observe that the previous discussion concerning the voltage regulator depicted in Fig. 4-20b applies only if the temperature of the zener diode does not vary under operating conditions. One method of compensating for the increasing or decreasing breakdown-diode resistance with increasing or decreasing temperature, respectively, is to include devices with a negative temperature coefficient in series with the breakdown diode. As an illustration, the configuration shown in Fig. 4-21 employs two forward-biased diodes (CR1 and CR2) in series with the breakdown diode (CR3). In turn, the total resistance of the three series-connected diodes remains constant over a wide temperature range. The net result is a constant voltage output despite temperature variation, input-voltage variation, and load-current variation over a considerable range. Two diodes are utilized in this example, because the temperature coefficient of resistance for each diode is one half that of the breakdown diode. Forward-biased diodes are employed because their voltage drop is very low. Observe that the audio circuit designer could utilize thermistors or other temperature-sensitive devices to achieve the same result, if he prefers.

### Voltage Stabilization by Breakdown Diode

With reference to Fig. 4-22, a breakdown (zener) diode is used in an amplifier circuit to provide collector-voltage stabilization. Observe that the current $I_2$ that is supplied by the battery divides into the breakdown diode

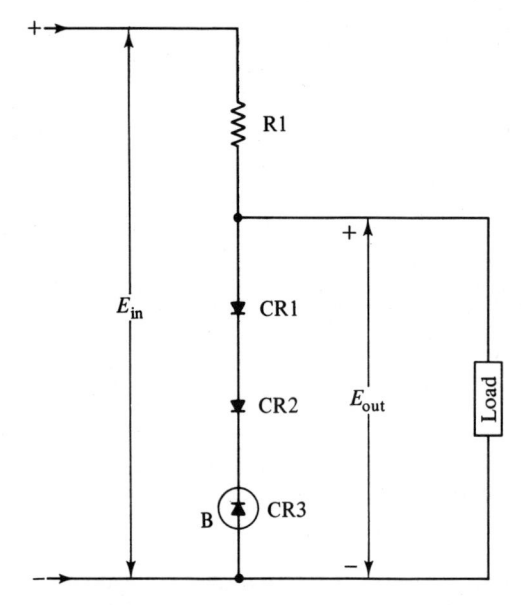

**Figure 4-21.** Breakdown diode temperature-compensated voltage regulator.

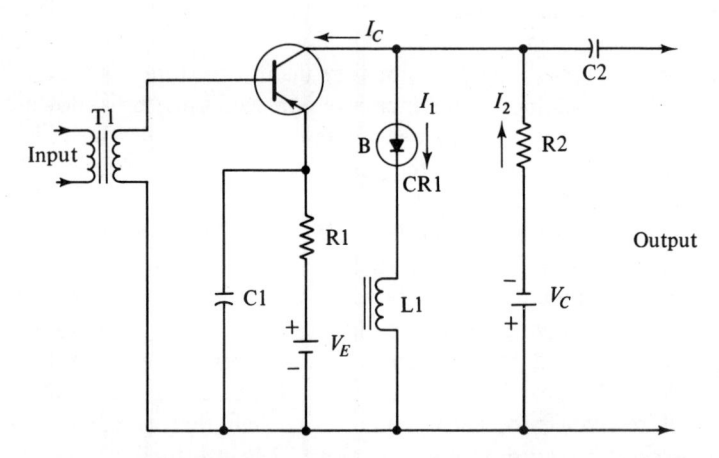

**Figure 4-22.** Breakdown diode used in an amplifier circuit to provide collector voltage stabilization.

current $I_1$ and the collector current $I_C$. When the dc collector current increases as a result of temperature rise, current $I_1$ decreases by the same amount, and the value of $I_2$ remains constant. The dc voltage drop across R2 also remains constant. In turn, the voltage applied to the collector (the battery voltage minus the drop across R2) also remains constant. This

## 4-6 Temperature Compensation by Breakdown (Zener) Diodes 149

analysis neglects the variation in resistance versus temperature of the breakdown diode CR1. However, to obviate this factor, the audio circuit designer can include two additional diodes, as in Fig. 4-21.

The ac resistance of a breakdown diode may vary from 5 to 1,000 Ω, depending upon the device design. To avoid shunting the collector load resistance by the low ac resistance of diode CR1, a high-impedance audio choke L1 is connected in series with the diode. Transformer T1 couples the ac input signal to the base–emitter circuit. Resistor R1 is the emitter swamping resistor. Battery $V_E$ supplies emitter–base bias voltage. Capacitor CI bypasses ac signal around resistor R1 and battery $V_E$. Capacitor C2 blocks dc voltage from and couples the ac signal to the following stage.

### Diode Surge Protection

If an excessive emitter–collector voltage occurs while the normally forward-biased base–emitter circuit is reverse biased, internal oscillation may occur that can destroy the transistor. This hazard may exist in transistor amplifiers that employ transformer-coupled input and output, as exemplified in Fig. 4-23. In the event that the signal input from the preceding stage is suddenly terminated, or if excessively strong noise signals drive the base–emitter circuit into a reverse-biased condition, the collector current cuts off rapidly. In turn, the magnetic field that surrounds the windings in T2 collapses quickly and generates a high emitter–collector counter emf that reverse biases the base–emitter junction. To forestall possible destruction of the transistor, the audio circuit designer may connect a junction diode between the base and emitter terminals to prevent the base–emitter junction from becoming reverse biased.

Observe that the voltage divider comprising R1 and R2 supplies a bias voltage for the base circuit that forward biases the base–emitter junction and reverse biases the junction diode (CR1). Under normal operating

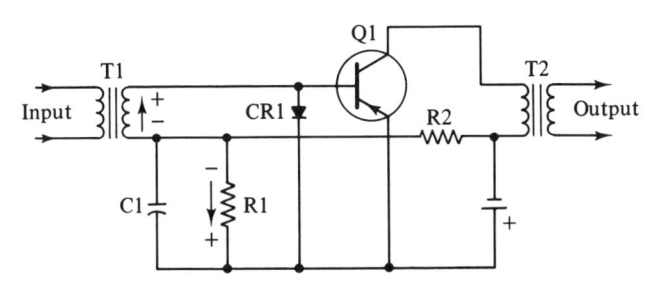

**Figure 4-23.** Junction diode action prevents the base–emitter junction of the transistor from becoming reverse biased.

conditions, CR1 is effectively an open circuit. In the event of a strong surge voltage of indicated polarity across the secondary of T1, CR1 will become forward biased and will conduct if the surge voltage exceeds the voltage across CR1. In its conducting state, only a very small voltage is dropped across CR1. This voltage drop is negligible and prevents the transistor's base–emitter junction from becoming reverse biased.

chapter five

# AUDIO POWER AMPLIFIERS

## 5-1 General Considerations

There is no sharp dividing line between small-signal audio amplifiers and audio power amplifiers. A high-power amplifier is illustrated in Fig. 5-1. Driver amplifiers occupy an intermediate position between a preamplifier and an output power amplifier, as depicted in Fig. 5-2. Although an audio power amplifier may operate in class A, nearly all commercial types operate either in class B or class AB. Some form of push–pull output configuration is generally utilized. Most audio power amplifiers are output transformer-

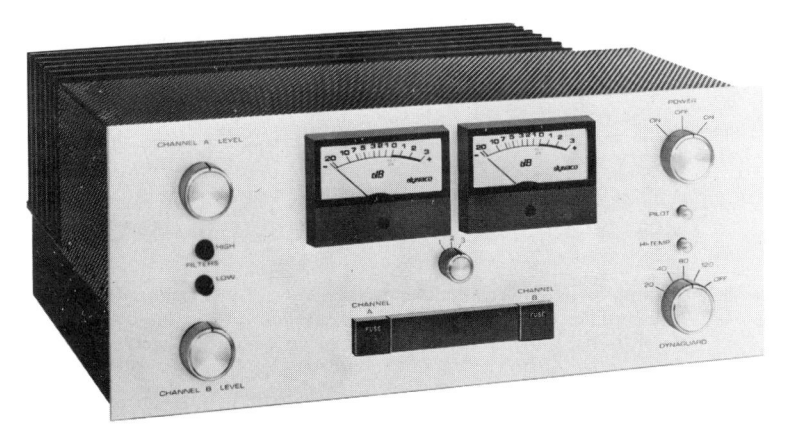

**Figure 5-1.** High-quality audio power amplifiers. (*Courtesy, Dynaco*)

**Figure 5-1.** (*Continued*)

(a)

(b)

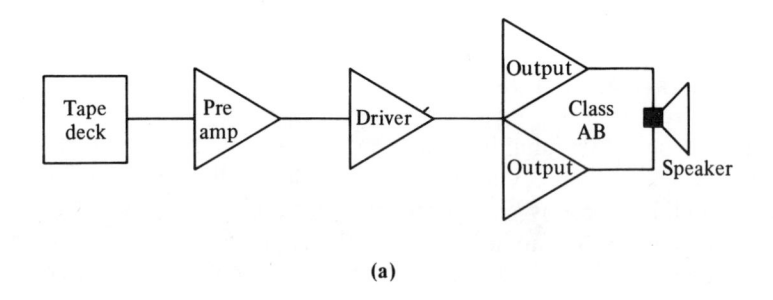

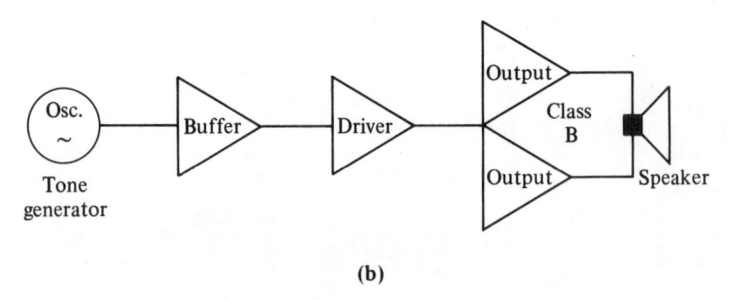

**Figure 5-2.** Examples of power-amplifier and driver arrangements: (a) tape deck, preamplifier, driver amplifier, and output power amplifier; (b) oscillator, buffer amplifier, driver amplifier, and output power amplifier.

less (OTL) designs. The arrangement depicted in Fig. 5-2a is representative of high-fidelity design; the arrangement shown in Fig. 5-2b is typical of electronic organs. Some form of negative feedback is always employed in an audio output configuration to minimize distortion. A power amplifier may be combined with a preamplifier in the same cabinet, or it may be enclosed in an individual cabinet. In the case of public-address (PA) systems, the power amplifier may be housed with the speaker in a common enclosure. There is an indicated trend to the mounting of high-fidelity power amplifiers in the same enclosure as the speaker system. This design facilitates voice-coil negative feedback and provides unusually good transient response.

## 5-2 Output-Transformerless Power Amplifiers

It was previously noted that audio transformers impose undesirable limitations on the design of high-fidelity amplifiers, particularly power amplifiers. Costly and bulky transformers are required for full high-fidelity frequency response. Phase shifts in the low- and high-frequency cutoff regions also impose difficulty in the design of negative-feedback loops. Therefore, output-transformerless (OTL) configurations are favored by hi-fi audio circuit designers. Three basic types of circuitry are employed. One approach utilizes a driver transformer to obtain phase inversion. In this function, the transformer is not labored as much as when it is utilized as an output component. In turn, its size and cost are comparatively low.

A second OTL configuration in general use is called the quasi-complementary circuit; it uses two identical output devices driven by two lower-powered complementary transistors that provide the equivalent of phase inversion. A third configuration, termed the fully complementary amplifier, employs a complementary pair of devices in its output section so that phase inversion is accomplished in the power output transistors, or in the combination of the output transistors and their drivers. We will consider the basic characteristics of these three design approaches.

## 5-3 Transformer Phase Inverters

A typical configuration that uses a transformer for phase inversion is shown in Fig. 5-3. Q1 and Q2 may be regarded as a two-stage voltage amplifier that drives the output transistors Q3 and Q4 through the driver transformer. Each transistor stage can be considered a power amplifier. In other words, Q1 develops a small amount of signal power to drive a larger transistor (Q2), which in turn develops sufficient power to drive the high-power output transistors Q3 and Q4. In theory, at least, Q3 and Q4 are larger devices than Q2, and Q2 is a larger device than Q1.

Observe that the input signal is capacitively coupled to Q1. However,

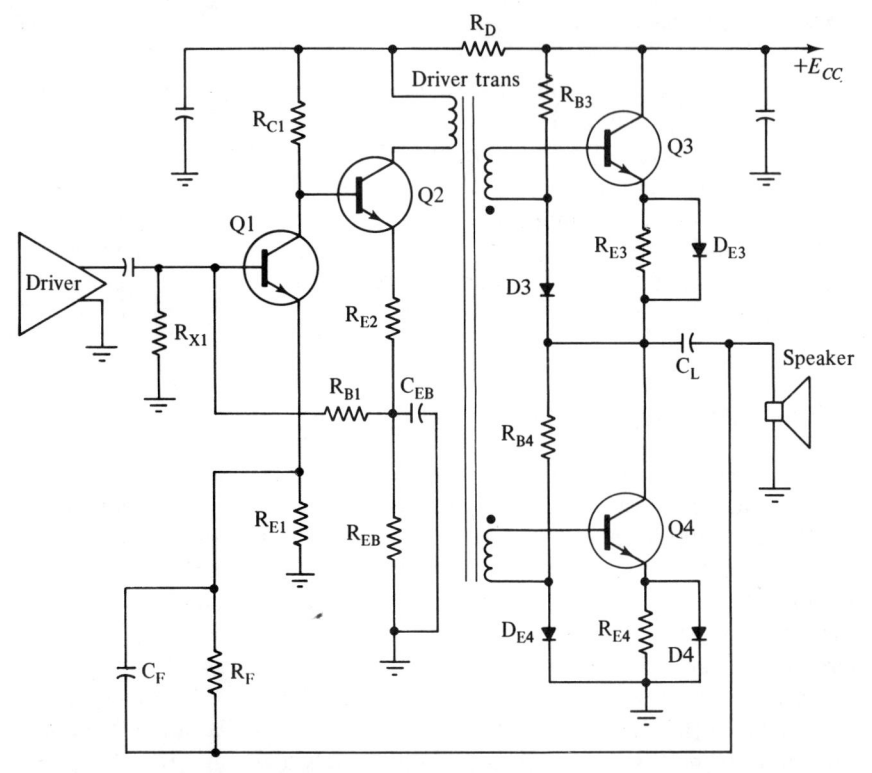

**Figure 5-3.** Driver transformer provides phase inversion for the push–pull output stage; may be used in PA equipment.

Q1 is direct coupled to Q2, and Q1 obtains its bias voltage from Q2's emitter circuit. Inasmuch as the dc feedback through $R_{B1}$ from the junction of $R_{E2}$ and $R_{EB}$ is appreciable, adequate temperature stabilization is realized. $R_{EB}$ in Q2's emitter circuit is bypassed by $C_{EB}$ to prevent any ac signal voltage from being fed back from this point along with the dc bias voltage. Next, the output from Q2 is fed to a transformer that has two identical secondary windings; the audio circuit designer often specifies bifilar windings. The phase relationship between these two windings is indicated by dots. Dots placed at the ends of the two windings indicate that these ends are in phase with respect to the unmarked ends.

When the instantaneous voltage in a cycle is such that the unmarked ends of the windings are positive with respect to the dotted ends, Q3 is forward biased, and this transistor conducts while Q4 is reverse biased and therefore cut off. In the following portion of the cycle, opposite polarity relations exist at the transistor bases; Q4 conducts while Q3 is cut off. Finally, the composite signal wave form is reconstituted across $R_L$. The impedance ratio of the transformer selected by the audio circuit designer is based upon considerations of presentation of an ideal load to the driver

## 5-3 Transformer Phase Inverters

transistor Q2. Typical designs utilize an impedance ratio of approximately 9 to 1. The audio circuit designer customarily optimizes the winding ratio for minimum distortion in his breadboard model.

If we assume that adequate transistors are chosen and are mounted on efficient heat sinks, the amount of audio signal power that the amplifier can deliver is based on the value of the supply voltage $E_{CC}$, upon the collector-to-emitter saturation voltage, and upon the voltage drops across emitter resistors $R_{E3}$ or $R_{E4}$. Output power is related to the output load resistance by the familiar terms $V^2_{rms}/R_L$ or $I^2_{rms}R_L$. Peak voltages and currents are 1.41 times greater than the rms values, and peak-to-peak voltages and currents are 2.83 times greater than the rms values. The power supply must have the capability of swinging the peak-to-peak output voltage across the load resistance, in addition to the peak-to-peak voltage swing that occurs across one of the emitter resistors $R_{E3}$ or $R_{E4}$.

Note that collector-to-emitter saturation voltage limits the swing of the voltage across the load. Inasmuch as two transistors are involved in this consideration, the sum of both of the saturation voltages at the peak of the collector current swing must be added to $V_{p-p} + I_{p-p}R_{E3}$ to estimate the minimum supply voltage that will be required if the amplifier is to deliver a specified amount of power. Restriction of the path of operation to within the linear region requires that the specified saturation voltage be multiplied by a factor of at least 3 before it is added to the other values in the relationship, in order to determine the minimum $E_{CC}$ supply voltage value that will be required if the amplifier is to deliver a specified amount of power without objectionable distortion.

We observe that diodes D3 and D4 in the output section are forward biased and are included by the designer to stabilize the quiescent current against variations of $V_{BE}$ versus temperature. Idling current is established by the voltage dropped across the diodes, in addition to the voltage dropped across the other resistors in the dc circuit. Resistors $R_{E3}$ and $R_{E4}$ in the emitter circuits are employed primarily to supply ac and dc negative feedback, and thereby reduce inherent distortion as well as contribute to temperature stabilization. As an auxiliary benefit, the emitter resistors assist in limiting the emitter current and thereby provide some measure of protection against overdissipation in the output transistors in case that $R_L$ is accidentally short-circuited. In class A amplifiers, approximately 0.5 to 1.5 V is developed across an emitter resistor; similar drops are desirable in class AB or class B operation on peak-current points of the signal cycle.

Diodes $D_{E3}$ and $D_{E4}$ may be omitted; however, they permit the use of higher-valued emitter resistors for improved temperature stabilization. In this situation, the diodes are required to bypass the resistors for obtaining large current swings. The amplified signal voltage is capacitively coupled to $R_L$; a large value of capacitance must be used to obtain low-frequency output. $R_F$ and $C_F$, in combination with $R_{E1}$, are the essential components of the feedback circuit. The circuit designer generally determines the value of $C_F$ experimentally for best transient response. Circuits with driver transformers are necessary when germanium-type output transistors are utilized. Germanium transistors have comparatively high leakage currents, and complete isolation of the output transistors by a driver transformer is desirable. Although it is possible to employ quasi-complementary circuitry with germanium transistors, it is best adapted to operation with silicon-type output transistors.

## 5-4 Quasi-Complementary Power Amplifiers

A complementary-symmetry power amplifier employs a pair of NPN and PNP output transistors to provide phase inversion and push–pull output with simplified and economical circuitry. The chief disadvantage of a complementary-symmetry output configuration is the relatively low power capability of the PNP transistor, compared with that of the NPN transistor. Therefore, circuit designers often utilize a quasi-complementary configuration, in which a low-current PNP transistor is directly coupled to a high-current NPN transistor in order to simulate the circuit action of a high-current PNP transistor. (See Chart 5-1.)

A basic configuration for the quasi-complementary arrangement is shown in Fig. 5-4. This circuitry is direct coupled throughout. The incoming signal voltage is amplified by Q1 and is then applied to the complementary pair Q2 and Q3. Over the positive portion of the signal cycle, the bases of the complementary pair are driven positive with respect to their emitters; the NPN-type transistor Q2 conducts, while the PNP-type transistor Q3 is turned off. On the next half-cycle, the transistor roles are reversed. Half-cycles of signal voltage are applied to the output transistors Q4 and Q5 after amplification by the complementary pair. Both portions of the signal cycle are fed to $R_L$ through $C_L$, and the input signal waveform is reconstituted across the load resistor. Negative feedback proceeds via the parallel combination of $R_F$–$C_F$.

Dc conditions in this configuration are such that half of the supply voltage must be present at point $E_{CC}/2$ in the diagram. The bias current is determined by resistors $R_{B1}$ and $R_{X1}$, and the bias current through Q1 essentially establishes the quiescent condition. The collector load for Q1 consists basically of $R_{B2} + R_{Y2} + D1 + D2$. These diodes are utilized to set and to maintain the idling current in the output circuit at the original quiescent value, despite temperature changes. However, the audio circuit designer may choose other types of temperature-sensitive devices. If temperature compensation is not essential, they may be replaced by resistors. Observe that capacitor C2 is included in a positive-feedback bootstrapping

## 5-4 Quasi-Complementary Power Amplifiers

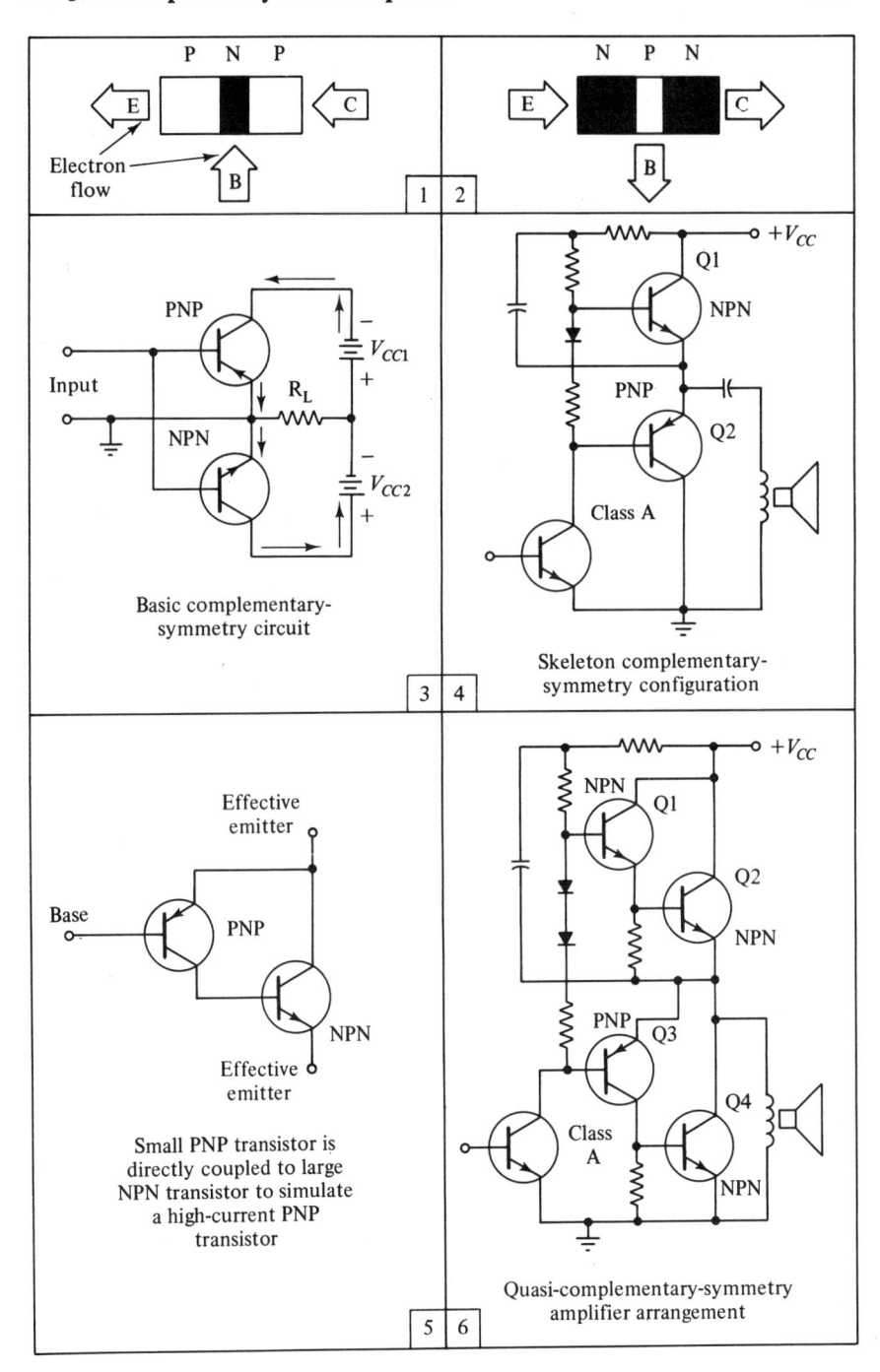

| | |
|---|---|
| **1.** P N P transistor (PNP) with Electron flow, E, B, C labels | **2.** N P N transistor (NPN) with E, B, C labels |
| **3.** Basic complementary-symmetry circuit | **4.** Skeleton complementary-symmetry configuration |
| **5.** Small PNP transistor is directly coupled to large NPN transistor to simulate a high-current PNP transistor (Effective emitter, Base, PNP, NPN) | **6.** Quasi-complementary-symmetry amplifier arrangement |

Chart 5-1.

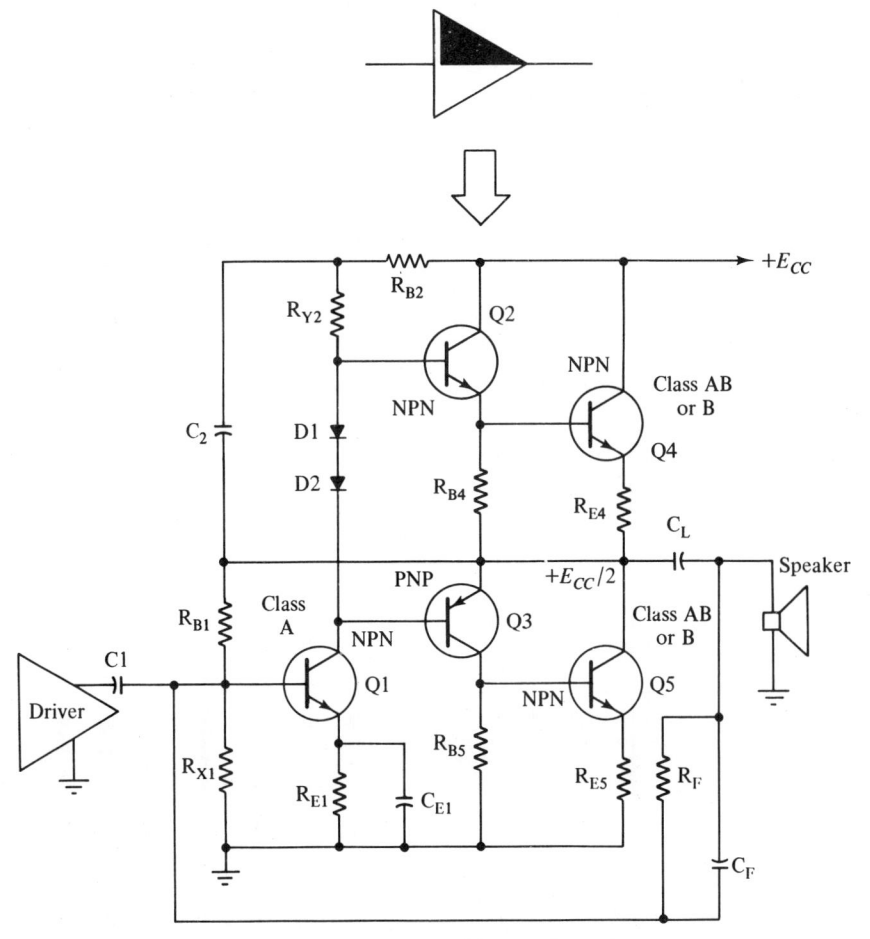

**Figure 5-4.** Basic quasi-complementary amplifier configuration.

circuit branch. Also, the resistors in the bias circuit for Q2 are split into components $R_{B2}$ and $R_{Y2}$, thereby providing a junction for connection of C2.

When the input signal has a comparatively high amplitude, the bias on the driver transistors tends to shift the operating point into class B, thereby introducing crossover distortion. To compensate for this shift in operating point, large amounts of feedback must be placed around the circuit. The gain must be appreciable to accommodate the amount of feedback that is required. Positive feedback via C2 increases the load impedance that the complementary pair presents to Q1 and increases the gain of the circuit. High-amplitude positive peaks in the signal tend to cut off Q2 by bringing the base and emitter terminals to $+V_{CC}$ potential. However, there

## 5-4 Quasi-Complementary Power Amplifiers

is a voltage across C2 as a result of its being charged while the circuit is idling. This voltage maintains the base at a positive potential with respect to the emitter, so that Q2 continues conducting over the complete signal cycle.

Design of the bootstrap circuit is comparatively simple. Inasmuch as $R_{B2}$ and $R_{Y2}$ are essentially connected across the load via C2, these resistors are chosen as large as possible consistent with the base current requirement of Q2. Equal values of resistance are utilized. Under quiescent conditions, $E_{CC}/2$ drops across the series circuit formed by $R_{B2}$ and $R_{Y2}$ as well as across the circuit formed by $R_{B2}$ and C2. Since $R_{B2} = R_{Y2}$, the voltage across C2 is one half of $E_{CC}/2$, or it is equal to $E_{CC}/4$. C2 charges to this voltage level and maintains a constant current through $R_{Y2}$ and Q2's base–emitter junction. C2 must be chosen sufficiently large that it can maintain its charge during low-frequency operation.

Another bootstrap circuit designed to operate without capacitor C2 and resistors $R_{B2}$ and $R_{Y2}$ is depicted in Fig. 5-5. Note that a resistor is connected from the junction of $C_L$ and $R_L$ to the base of Q2. Thus $C_L$ doubles as a bootstrap capacitor in addition to coupling the signal to the output load resistor or speaker. Load resistor $R_L$ is connected to $+E_{CC}$, which is at ac ground potential. Observe that this circuit has a drawback, in

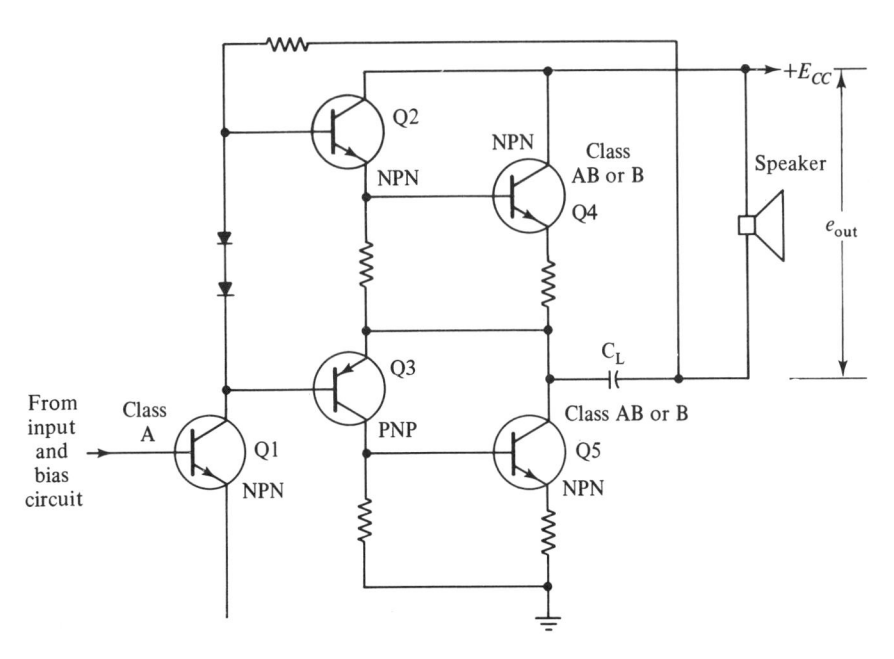

**Figure 5-5.** This arrangement eliminates the bootstrap capacitor by combining its function with that of the output blocking capacitor.

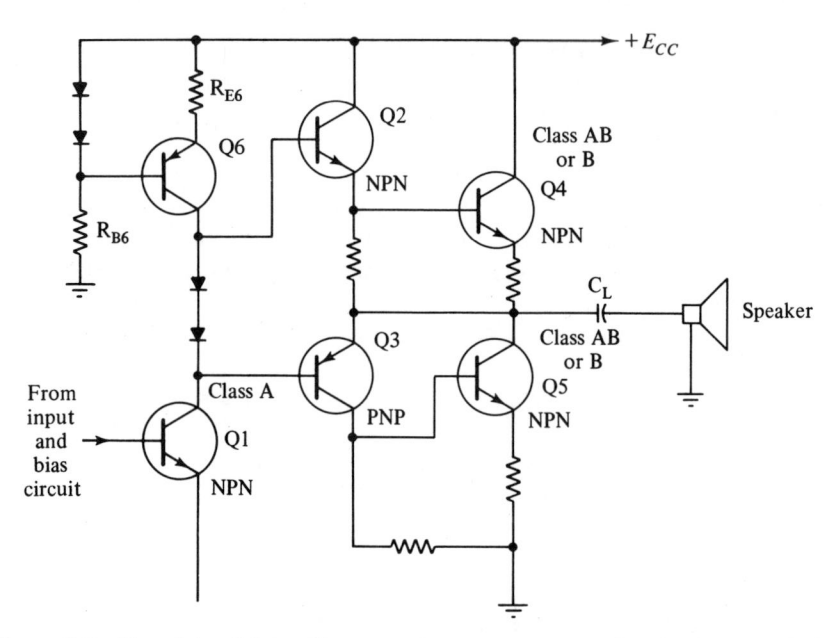

**Figure 5-6.** Transistor Q6 provides a constant-current source for the drivers, eliminates a bootstrap capacitor, and reduces low-frequency distortion.

that the dc base current for the drivers flows through $R_L$. If this current has a small value, it will not adversely affect a loudspeaker used as load $R_L$. A constant-current source for the bases of the complementary drivers can eliminate the requirement of a bootstrap capacitor. Refer to Fig. 5-6. This arrangement applies a constant current to the drivers and presents a high impedance to Q1, the voltage amplifier transistor. The voltage drop between the base of Q6 and $+E_{CC}$ is made as small as possible in order to avoid limiting the output voltage swing. Accordingly, low forward-voltage dropping diodes are ordinarily chosen for the constant-current circuit, instead of comparatively high voltage zener diodes. Advantages of this configuration include lower distortion in low-frequency operation and more symmetrical clipping of peaks on both half-cycles.

## 5-5 Directly Coupled Load

A large electrolytic capacitor is utilized in the foregoing configurations to couple the load to the output transistors. This component has some disadvantages, among which its nonlinear characteristic, low-frequency rolloff in combination with $R_L$, and the corner frequency created by the rolloff, come into consideration. Instability may be encountered when negative

## 5-5 Directly Coupled Load

feedback is applied around the circuit. Moreover, this coupling capacitor must be charged through the output transistors. This can lead to transistor damage in the event that the power–time product happens to exceed the transistor's maximum rating. Refer to Fig. 5-4. One end of $R_L$ is connected to ground. While the circuit is idling, the other end of $R_L$ must be at the same ground potential in order to avoid any dc current flow through the loudspeaker load. A coupling capacitor fulfills this requirement.

In the absence of a coupling capacitor $C_L$, dc current flow must be eliminated by placing the junction of Q4 and Q5 at a zero potential with respect to ground while the circuit idles. To accomplish this circuit action, a positive voltage with respect to ground, $+E_{CC}$, is applied to Q4's collector and a similar negative voltage, $-E_{CC}$, is applied at Q5's emitter or, more precisely, at the lower end of resistor $R_{E5}$. If both of these transistors conduct equal values of current during the idling period, there is 0 V at the junction of the two devices to which the load is connected (or at the junction of Q5 and the lower end of $R_{E4}$). With a signal voltage applied, the positive-going portion of the wave form swings the signal voltage across $R_L$ from zero toward $+E_{CC}$, while the negative-going portion of the wave form swings the signal voltage from zero toward $-E_{CC}$.

One problem is to maintain a constant quiescent current at all times, so that the voltage across $R_L$ remains zero. However, the current through Q1 will drift as the temperature changes. Any change in Q1's collector current will upset the balance at the output more than will current drift in any transistor farther up the chain. Current drift due to the driver and output stages is minimized by maintaining the upper and corresponding lower devices at equal temperatures on heat sinks or in free air. In turn, drift in one half of the driver and output circuit is balanced out by the drift in the other half of the configuration. Note that Q1 has no corresponding transistor to overcome or balance collector-current changes resulting from drift. Therefore, a balancing device must be added to the circuit.

A differential amplifier circuit employs two transistors. The audio-circuit designer should mount the two transistors in proximity so that temperature variations affect each equally. A basic configuration is depicted in Fig. 5-7. Q1 and Q2 operate in a differential circuit that prevents collector-current changes resulting from temperature variation. This differential pair drives the second differential pair Q3 and Q4. When a signal voltage is applied at the input of Q1, it appears in amplified form at the collectors of Q1 and Q2. The input signal is in phase with the signal at Q2's collector, whereas the input signal is out of phase with the signal at Q1's collector. As the signal is again amplified by Q3 and Q4, it is inverted once more in passage through each transistor. The collectors of both transistors are driven to $-E_{CC}$ by the signal voltage. Negative drive voltage is provided at Q3's collector to the lower pair of output devices, Q7 and Q9.

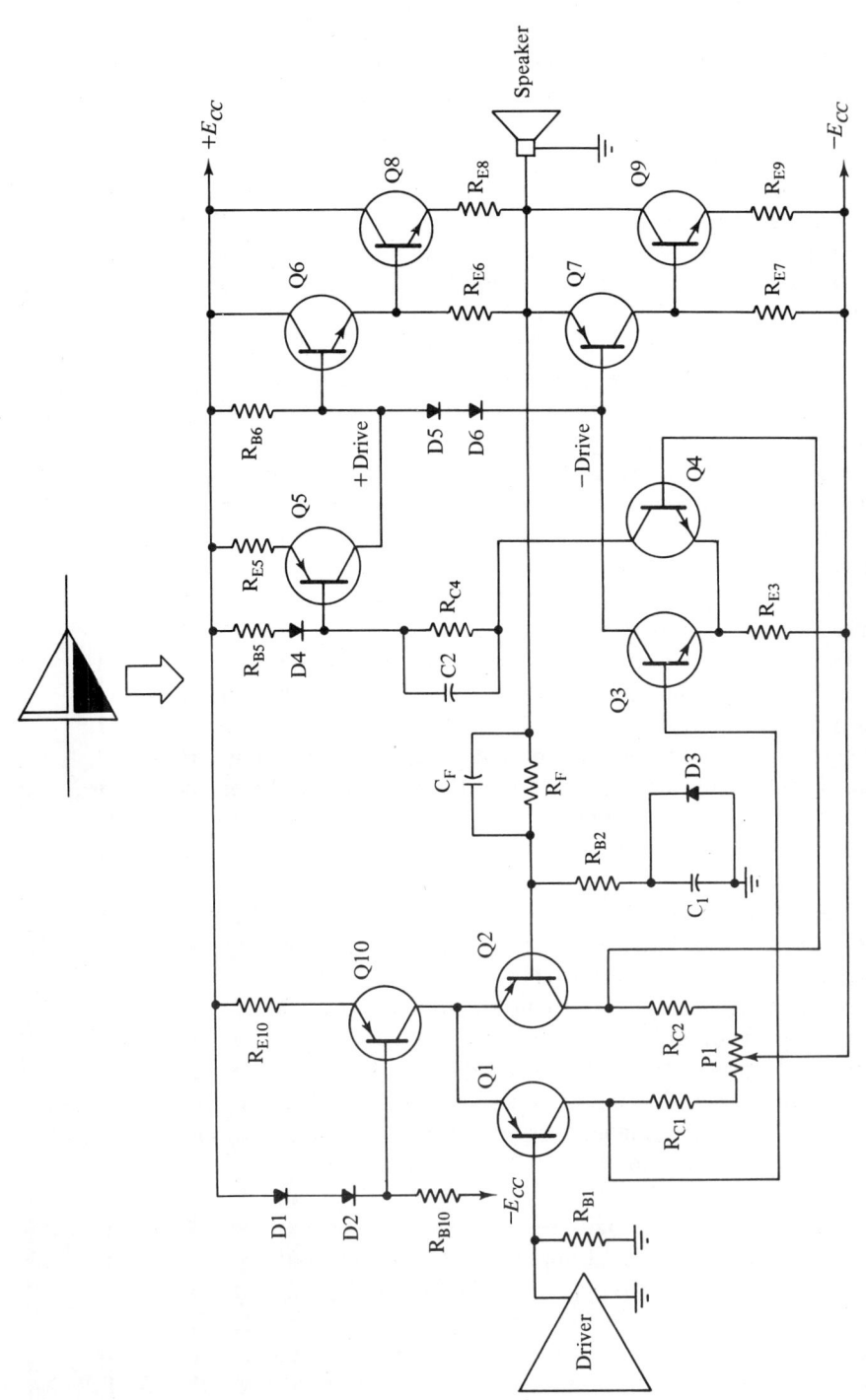

**Figure 5-7.** Differential-amplifier input circuit minimizes changes in quiescent current relations versus temperature.

## 5-5 Directly Coupled Load

Next, from Q4 the signal voltage is fed to the base of Q5; this transistor operates as a unity-gain amplifier, and it inverts the signal phase. The applied signal voltage drives the collector of Q5 to $+E_{CC}$, and thereby provides the positive drive voltage. Phase reversal through Q5 makes the positive drive signal identical in phase with the negative drive signal at Q3. In turn, two signals with identical phase characteristics are fed to the quasi-complementary output circuit comprising Q6 through Q9. Suitable phase relationships are achieved here to provide a reconstituted and amplified version of the input signal. The positive drive causes the upper pair of devices to swing to $+E_{CC}$, while the lower pair of devices swings from $-E_{CC}$ as a result of the negative drive. Equal swings of both halves of the output signal waveform to $+E_{CC}$ and to $-E_{CC}$ makes bootstrapping unnecessary.

Negative feedback is applied from the output of the amplifier via $R_F$ and $C_F$. Both ac and dc feedback voltage are developed across $R_{B2}$. This resistor is bypassed to ground by C1 to provide a return for the ac feedback voltage, and diode D3 is connected to ground to provide a dc feedback voltage return path. The audio circuit designer should not assume that this circuit is in finished form. It is unlikely to operate satisfactorily unless all component values and the tolerances on both components and devices are carefully assigned in view of the stipulated performance specifications. Note that instability is likely to be encountered unless the frequency rolloff is carefully controlled at both the low- and the high-frequency ends of the audio band. It is instructive to analyze various sections of the circuitry, as follows.

Refer to Fig. 5-7. Q10 is a constant-current amplifier transistor; its circuit is designed to establish and to maintain the sum of the idling currents through Q1 and Q2. If the circuit designer employs silicon devices, the voltages across the base–emitter junction and across D2 are identical. Similarly, the voltages across D1 and $R_{E10}$ are identical and have a value of approximately 0.7 V. This is in the range of the normal voltage drop across a silicon diode. Suppose that the sum of the idling currents through Q1 and Q2 is adjusted to a value of 2 mA. In turn, a current flow of 2 mA occurs through $R_{E10}$. In turn, the resistor will have a value of 0.7/0.002, or 350 Ω. Note that $R_{B10}$ is utilized to establish a current level through and consequently to fix the voltage dropped across D1 and D2.

To determine the idling current for D1 and D2, first note the maximum value of current required in the output load at the signal peak. This current value is divided by the product of the current gains of all the stages, with the exception of the first stage. In turn, the designer knows the minimum value of collector current required from each of the two output devices. Observe that if Q3 and Q4 are to be capable of swinging almost to $-E_{CC}$, then the voltage across $R_{E3}$ must be small. Assume that this range is 1.5 to 2 V. In turn, this value of voltage must also be present at the collectors of Q1 and Q2. During the idling intervals, a current of 1 mA is to flow through each collector. Accordingly, a simple Ohm's-law calculation will assist the designer in determination of the resistance values for the collector circuits. A portion of each collector resistance is subtracted from $R_{C1}$ and from $R_{C2}$ to provide for the resistance of potentiometer P1. This control serves to balance the relative quiescent currents through the transistors; it is adjusted for 0 V across $R_L$ while the circuit is idling. As a result of dc negative feedback, this adjustment is stabilized.

It follows from our previous discussion that the positive drive voltage must be equal to the negative drive voltage. In turn, $R_{B5}$ is assigned an equal value to $R_{E5}$. The voltage drops across D4 and the base–emitter junction are equal. Consequently, the currents through $R_{B5}$ and $R_{E5}$ are equal, threby producing a negative drive current in the collector circuit equal to the current flow through $R_{B5}$. Since the current through $R_{B5}$ is almost identical to the collector current that flows through Q3 and Q4, the negative drive voltage is equal to the positive drive voltage. Power that is available at the collector of Q4 and that is not required at the base of Q5 is dissipated by $R_{C4}$. The level of quiescent idling current through the output devices is set by action of D5 and D6. If the audio circuit designer desires adjustment flexibility, one of the diodes may be replaced by a potentiometer. Better temperature compensation can be realized if the diodes are replaced with transistor constant-current circuits, as depicted in Fig. 5-8.

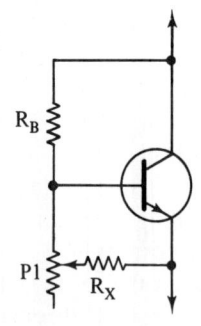

**Figure 5-8.** Transistor adjustable constant-current circuit.

Refer to Fig. 5-9. This is a simplified version of the circuit that was shown in Fig. 5-7. In this variation, the differential output is taken from one transistor of the pair Q1–Q2. $R_{E1}$ is chosen sufficiently large that a constant-current source is not required; this eliminates the cost of one transistor. The amplified signal voltage is applied to Q3 and is then fed to the quasi-complementary circuit comprising Q4 through Q7. Bootstrapping is provided by capacitor C1 so that Q4 can swing to saturation. Independent

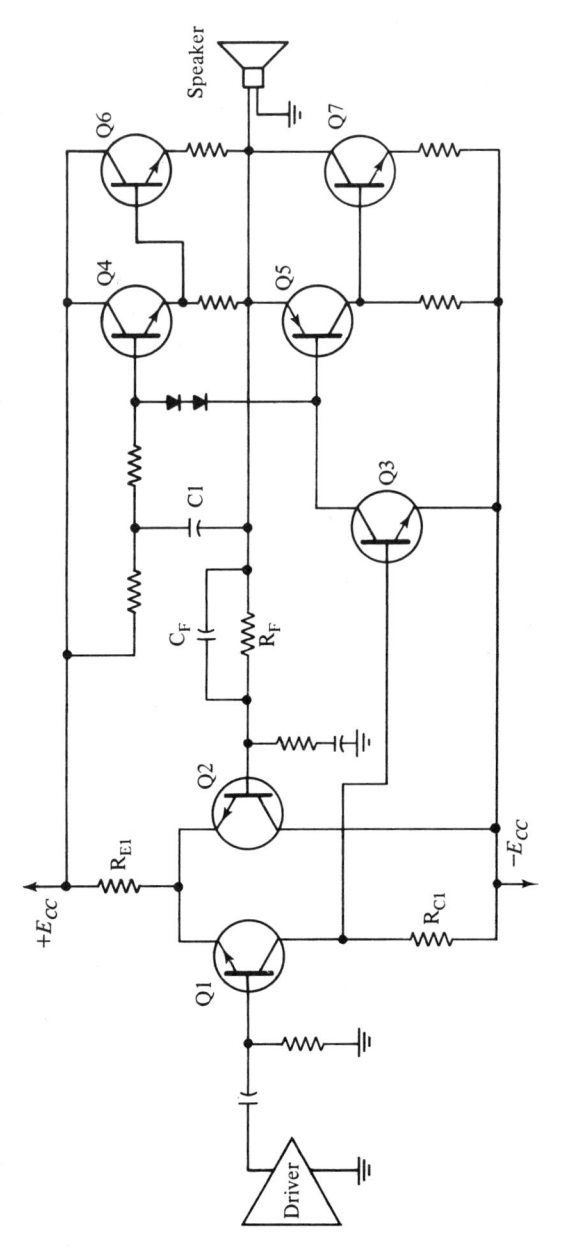

**Figure 5-9.** Simplified version of the circuit depicted in Fig. 5-7.

drivers with the capability of swinging the signal voltage to the limits of the power-supply voltage function to make bootstrapping unnecessary in the version of Fig. 5-7.

## 5-6 Full Complementary Output Arrangement

Analysis of the quasi-complementary configuration reveals that the upper two transistors operate as a Darlington pair, whereas the lower two transistors operate as a complementary beta-multiplier pair. Although these are quite similar arrangements, less distortion will result if the audio circuit designer employs identical pairs. He may choose either a complementary pair or a Darlington pair, as exemplified in Fig. 5-10. Either of these circuits can be substituted directly into any of the quasi-complementary arrangements that were illustrated. Either of the circuits can be employed to re-

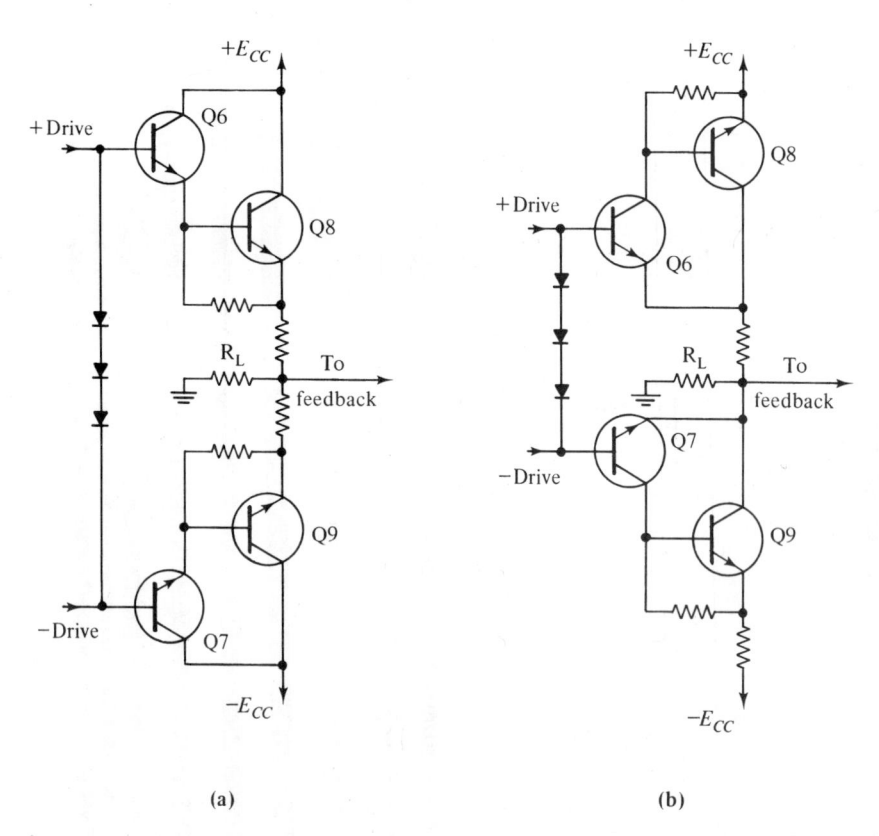

**Figure 5-10.** Symmetrical output circuit arrangements: (a) dual Darlington; (b) dual beta multiplier.

## 5-6 Full Complementary Output Arrangement

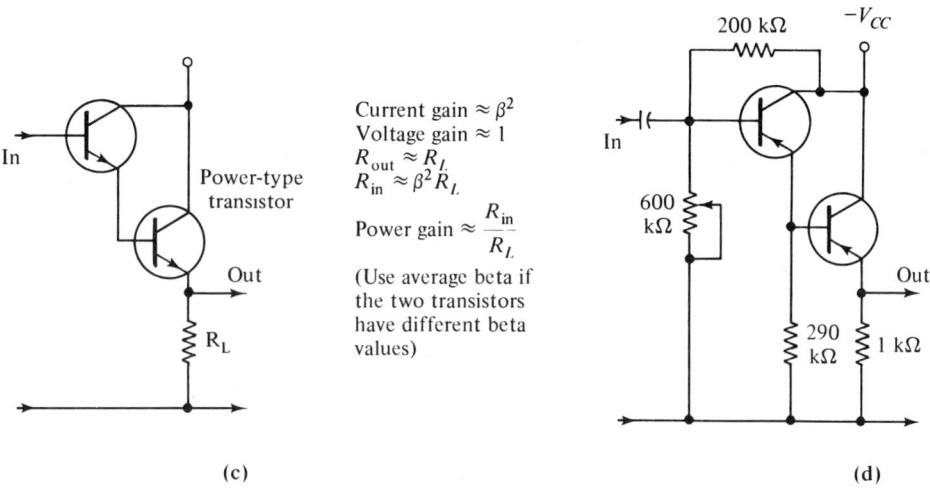

Current gain $\approx \beta^2$
Voltage gain $\approx 1$
$R_{out} \approx R_L$
$R_{in} \approx \beta^2 R_L$

Power gain $\approx \dfrac{R_{in}}{R_L}$

(Use average beta if the two transistors have different beta values)

(c)        (d)

**Figure 5-10.** (*Continued*) (c) basic Darlington compound (Darlington) circuit; (d) practical arrangement with bias resistors.

place transistors Q6 through Q8 in the configuration of Fig. 5-7. The same general design principles are observed that have been described.

Various circuit designers prefer the *complementary-symmetry* type of audio power amplifier. It offers the advantages of reduced circuit complexity, eliminates the need for a separate phase-inverter stage, and provides extended frequency response as a result of reduced common-mode conduction. A skeleton circuit diagram for a complementary-symmetry amplifier is shown in Fig. 5-11. Note that Q1 and Q2 operate in the CC mode. Each transistor conducts over one half of an input cycle, because Q1 is a PNP-type, whereas Q2 is an NPN-type transistor. Resultant output circuit action can be followed by analysis of the simplified arrangement depicted in Fig. 5-12. Note that the internal emitter–collector circuit of Q1 is represented by the variable resistor R1, and that of Q2 is represented by the variable resistor R2. While the arm of one variable resistor is in its "off" position, the other resistor arm varies through its range. Next, these relations reverse. Observe that Q1 and Q2 must have closely matched characteristics or the output waveform will contain amplitude distortion.

Power amplifiers usually require transistor operation at power levels that are near thermal runaway conditions. This hazard is aggravated by biasing networks that have marginal stability, but which may be chosen by the circuit designer because of increased operating efficiency. Inasmuch as thermal runaway in a power stage is almost certain to damage or destroy the transistors, the audio circuit designer must give careful attention to worst-case principles in order to eliminate, or at least to minimize, the possibility of thermal runaway. Worst-case conditions include the onset of

**168**  Audio Power Amplifiers

**(a)**

Typical transfer characteristic (left plot): Collector, mA vs Base to emitter, V

Typical transfer characteristic (right plot): Collector, mA vs Base, mA

**(b)**

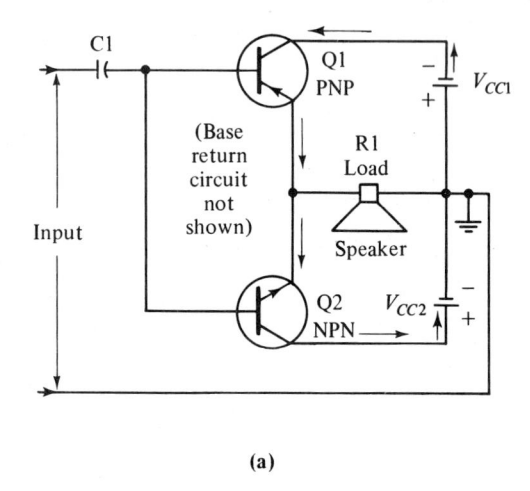

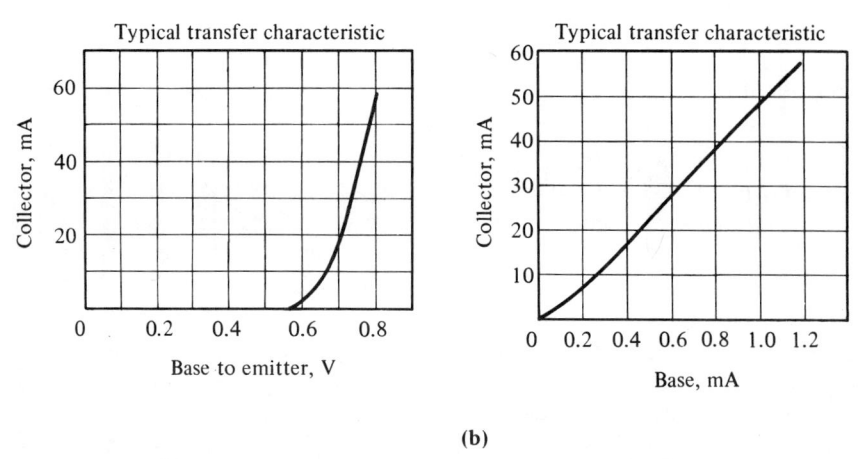

**Figure 5-11.** Zero-bias complementary-symmetry configuration: (a) skeleton circuit diagram; (b) typical power-transistor transfer characteristics.

indefinite increase in current gain ($h_{fe}$), zero base–emitter voltage, minimal load impedance, and saturation current ($I_{co}$) at a maximum value. In a class B amplifier, the maximum transistor power dissipation occurs at the time when the signal power output is 40 percent of its maximum value. At this time, the power dissipated by each transistor is 20 percent of the maximum power output. For purposes of comparison, in a class A amplifier, maximum power dissipation occurs in the absence of an input signal. The arrangement shown in Fig. 5-11 is one of the basic OTL types of audio-

## 5-6 Full Complementary Output Arrangement

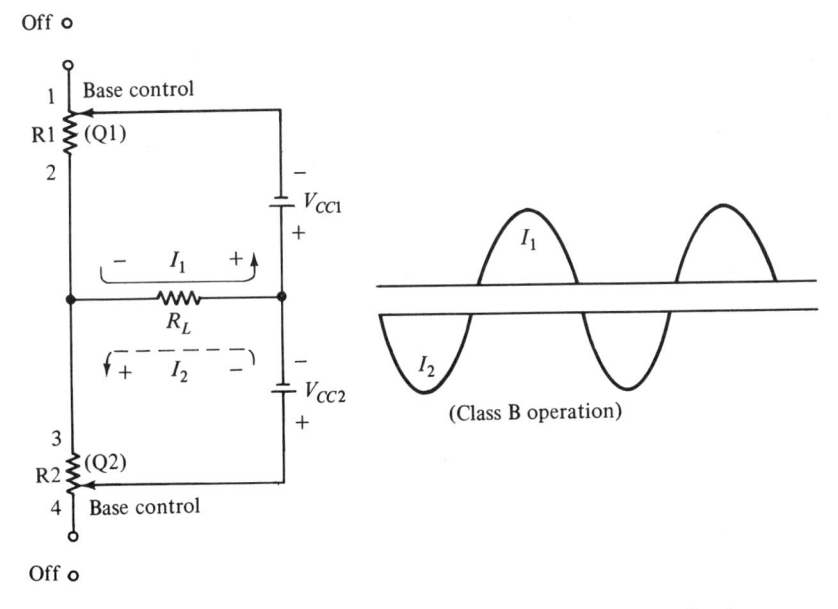

**Figure 5-12.** Simplified version of a complementary-symmetry output circuit.

output configurations. Most complementary-symmetry amplifiers are operated in class AB. Sufficient forward bias is applied to the transistors to avoid crossover distortion. With reference to Fig. 5-13, a small forward bias is applied; this is an emitter-follower configuration in which $R_L$, R2, and R3 provide forward bias for Q1 and Q2. Bias resistor R1 is connected in series with the base–emitter junctions of Q1 and Q2. If the two junctions have equal resistance values, the voltage drop across each junction will be one half of the total voltage drop across R1. In practice, the value of R1 is very small, and its unbalancing effect on the input signal to Q2 is negligible. Optimum class AB operation will be obtained only if the characteristics of Q1 and Q2 are reasonably well matched. Mismatch of these transistors can result in a combination of crossover distortion and stretching distortion. If the transistors are well matched and too much forward bias is employed, both of the transistors will develop stretching distortion.

### Cascaded Complementary-Symmetry Audio Power Amplifiers

Many audio power amplifiers are designed with complementary-symmetry circuitry in the common-emitter mode. Direct coupling is generally utilized, with an OTL configuration as exemplified in Fig. 5-14. One CE comple-

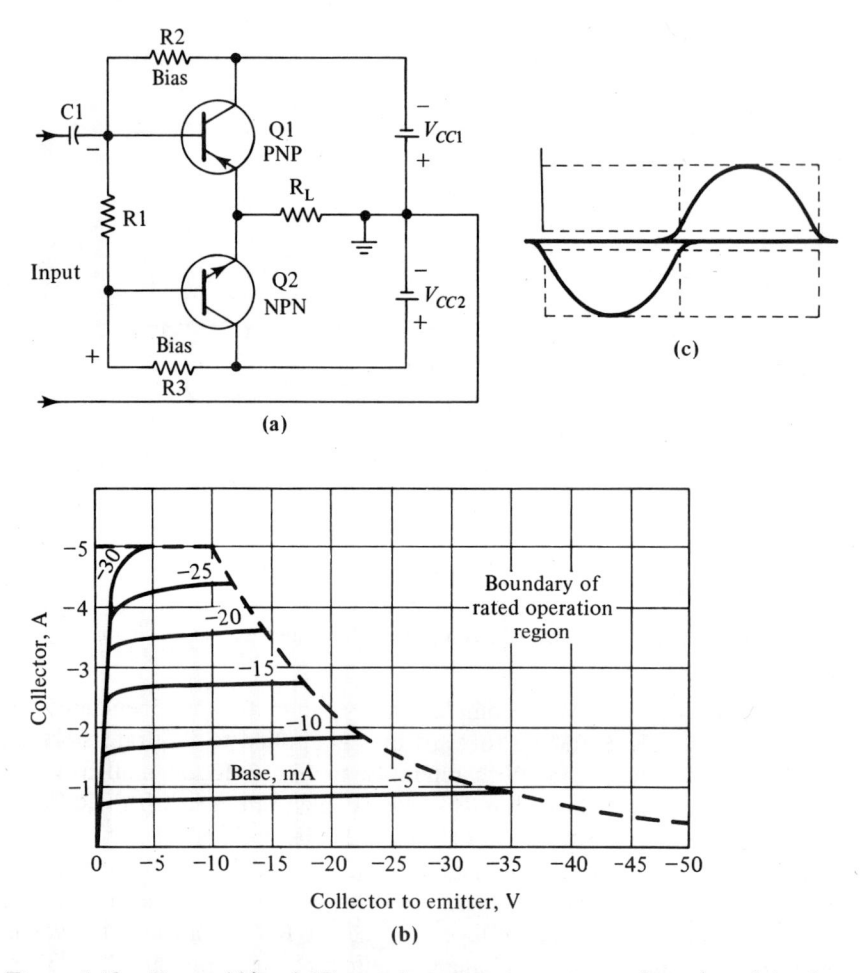

**Figure 5-13.** Forward-biased CE complementary-symmetry configuration: (a) schematic diagram; (b) maximum power dissipation rating for a typical power transistor; (c) small forward bias minimizes crossover distortion.

mentary-symmetry stage (Q3–Q4) is directly driven by another CE complementary-symmetry stage (Q1–Q2). A single-ended input signal is utilized. When the input signal goes positive, Q1 conducts and Q2 remains nonconducting. Because of the 180-degree phase reversal that occurs from input to output in the CE configuration, Q1's collector is negative going; this causes Q3 to conduct, and Q3's collector is positive going. When the input signal goes negative, Q2 conducts and its collector is positive going. This causes Q4 to conduct and its collector is negative going. Transistors Q1 and Q3 are nonconducting during this interval. Battery $V_{EE1}$ supplies

## 5-6 Full Complementary Output Arrangement

the required biasing voltages for Q1 and Q3; battery $V_{EE2}$ supplies the required biasing voltages for Q2 and Q4.

Observe that the base–emitter junction of Q3 is connected in series with the collector–emitter circuit of Q1 and $V_{EE1}$. Accordingly, Q3's emitter is positive with respect to its base (forward bias), and Q1's collector is positive with respect to its emitter, as required for electron flow through Q1. Similar circuit action occurs in the arrangement of Fig. 5-14b, with a different supply-voltage location; the audio signal developed across the speaker provides negative feedback in the input branch of Q1 and Q2, thereby providing a high value of input impedance. Transistors Q3 and Q4 are connected in a CE configuration to match the output resistance of Q1 and Q2. As in amplifier stages with a single transistor, the CE configuration in the cascaded complementary-symmetry arrangement provides higher power gain to a low-impedance load.

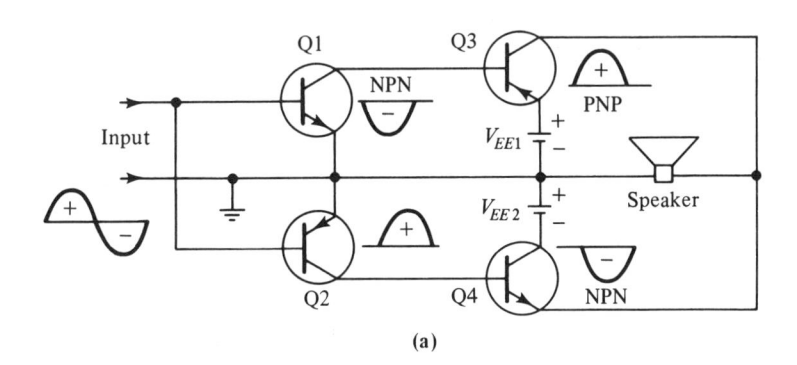

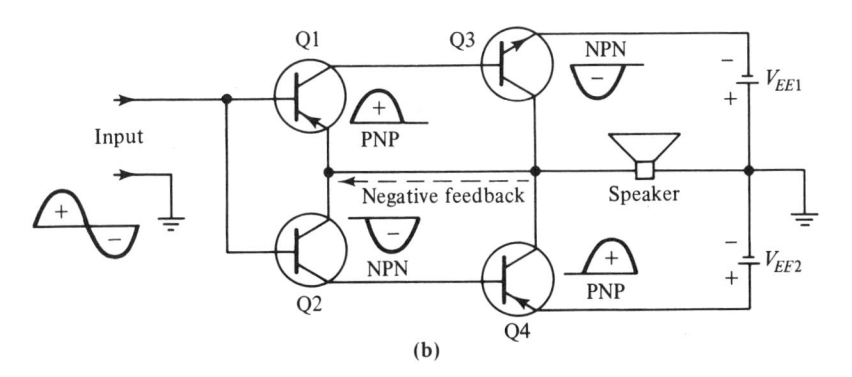

**Figure 5-14.** Examples of direct-coupled complementary-symmetry stages.

## Compound-Connected Complementary-Symmetry Configuration

The current gain, voltage gain, and power gain of a transistor are directly proportional to its short-circuit forward-current amplification factor. This factor is defined as the ratio of the output current to the input current, with the output load equal to zero (short circuited). To obtain maximum gain in an amplifier stage, it is necessary to utilize a transistor that has a high value of short-circuit forward-amplification factor. Most power transistors have an $\alpha_{fb}$ value ranging from 0.940 to 0.985, with an average value of 0.960. Note, however, that no matter what the value of $\alpha_{fb}$ may be, the amplification factor will decrease as the emitter current increases. Hence, when an amplifier stage employs a single transistor, there is inevitably a nonlinear relation of emitter current to collector current. This nonlinearity becomes most prominent at high current levels (peak power output).

This nonlinear relationship results in a reduction of the current amplification factor for a single transistor at high values of emitter current. In a power amplifier that draws heavy emitter current and which is operated near the maximum rated output of the transistor, this variation is aggravated. On the other hand, if two transistors are compound connected, as exemplified in Fig. 5-15, this nonlinear relationship can be minimized. The dashed lines in the diagram enclose the pair of compound-connected transistors. Observe that the base of Q1 is connected to the emitter of Q2, and that the two collectors are connected together. Both of the transistors operate in the CB configuration. The current gain for the compound-

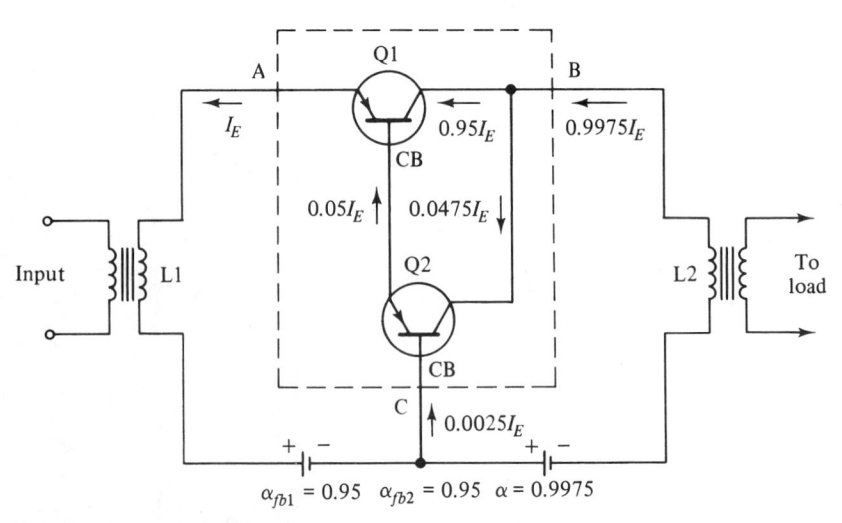

**Figure 5-15.** Flow diagram for compound-connected transistors.

## 5-6 Full Complementary Output Arrangement

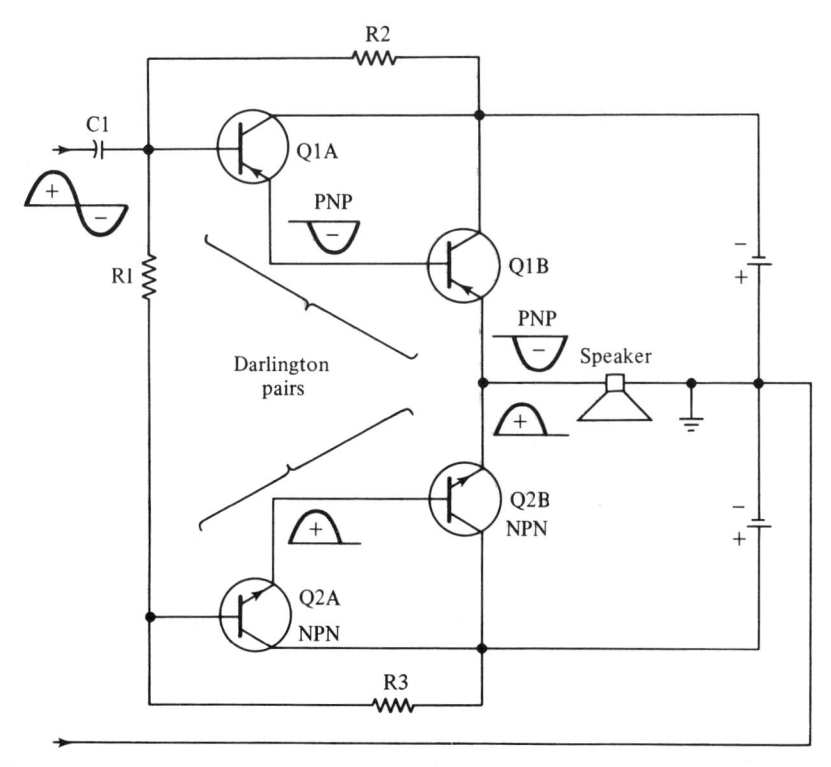

**Figure 5-16.** Complementary-symmetry compound-connected (Darlington) output amplifier arrangement.

connected transistors is greater than for a single transistor, in addition to providing greatly improved linearity of operation. If we assume that each transistor has a current-gain value of 0.95, their combined current-gain value becomes 0.9975 in the compound connection. This increase corresponds to a beta increase from 19 to 399.

It is not essential that the transistors in Fig. 5-15 have equal current-amplification factors. Also, compound-connected transistors are employed to advantage in single-ended amplifiers, in conventional push–pull amplifiers, and in complementary-symmetry amplifiers. An example of compound-connected (Darlington) transistors in a complementary-symmetry power-amplifier arrangement is shown in Fig. 5-16. This configuration is basically similar to the complementary-symmetry arrangement depicted in Fig. 5-13. However, Q1 is replaced by the compound connection of Q1A and Q1B in Fig. 5-16. Similarly, Q2 is replaced by the compound connection of Q2A and Q2B. Since the pairs of transistors are connected as Darlington pairs, this arrangement is also called a Darlington-pair complementary-symmetry configuration. Inasmuch as the transistors operate

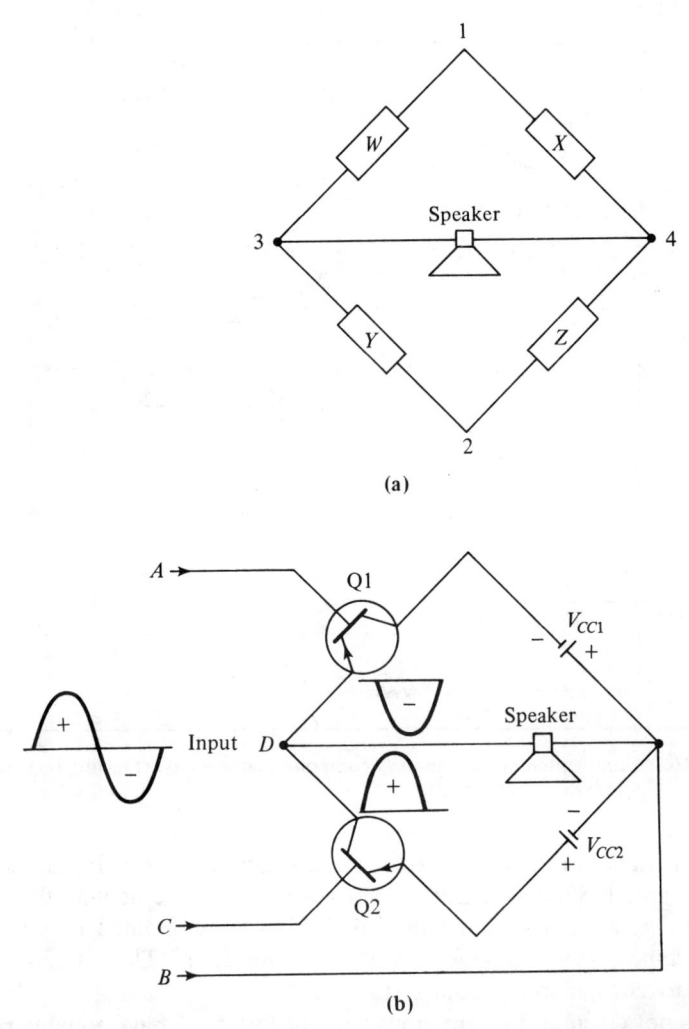

**Figure 5-17.** Basic bridge arrangement, and a bridge with two transistors and two batteries: (a) basic bridge circuit; (b) transistors and batteries operate in bridge arms.

basically in the CC mode, a large amount of negative feedback occurs, and the percentage of distortion is very low.

### Complementary-Symmetry Bridge Arrangement

Audio power amplifiers are also configured in the complementary-symmetry bridge arrangement, as exemplified in Fig. 5-17. The arms of the bridge, $W$, $X$, $Y$, $Z$, can be arranged for balance, whether the circuit ele-

## 5-6 Full Complementary Output Arrangement

ments are resistors, capacitors, transistors, or batteries. Bridge balance occurs when, regardless of the voltage applied at points 1 and 2, the voltage drops across the bridge arms are such that zero voltage is developed across points 3 and 4. In other words, no current flows through the speaker. Thus the resistors in the basic bridge circuit can be replaced by transistors Q1 and Q2, and by batteries $V_{CC1}$ and $V_{CC2}$, as depicted in Fig. 5-17b. If, in turn, the transistors are biased to draw equal emitter currents (or to draw no emitter–collector current) under quiescent (idling) conditions, the bridge is balanced and zero current flows through the speaker. On the other hand, if an input signal causes either of the transistors to conduct more current than the other, the bridge becomes unbalanced and current flows through the speaker. Sine-wave signals applied to points $A–B$ and $C–D$ that are 180 degrees out of phase with each other will result in an amplified sine-wave signal across the speaker. If the transistors are operated in class A or B, no dc current will flow through the speaker.

It is undesirable to have a dc current flow through a speaker voice coil because this produces cone offset, which tends to distort the reproduced sound. Observe that, if a conventional push–pull power amplifier were utilized, the speaker voice coil would need to be center tapped, and only one half of the coil could be used for each half-cycle of the input signal. Such center-tapped operation entails reduced efficiency in conversion of electrical energy into sound energy. Observe the bridge configuration depicted in Fig. 5-18. All four arms of the bridge consist of transistors. Q1 and Q3 are PNP types, whereas Q2 and Q4 are NPN types. This configuration has an advantage in that no part of the input circuit or of the output circuit need be operated at ground potential. Moreover, a complementary-symmetry bridge amplifier can be energized from a single-ended driver.

Assume under idling conditions that all the transistors in Fig. 5-18 are zero biased and that they draw no current (class B operation). If the transistors draw no current, there is no completed circuit across $V_{CC}$ and no current flows through the speaker. Next, if an input signal drives point $A$ negative with respect to point $B$, Q1 and Q4 will become forward biased, and Q2 and Q3 will become reverse biased. Therefore, Q1 and Q4 will conduct; electrons will flow from the negative battery terminal through Q1 collector-to-emitter, through the speaker, through Q4 emitter-to-collector, and back to the positive battery terminal. This path for current flow is indicated by the solid-line arrows, and a voltage of the indicated polarity is dropped across the speaker.

Next, if an input signal causes point $A$ to become positive with respect to point $B$, Q3 and Q2 will conduct, and Q1 and Q4 will be reverse biased. In turn, the path of electron current is indicated by the dashed-line arrows. A voltage of the indicated polarity is dropped across the speaker. Assume

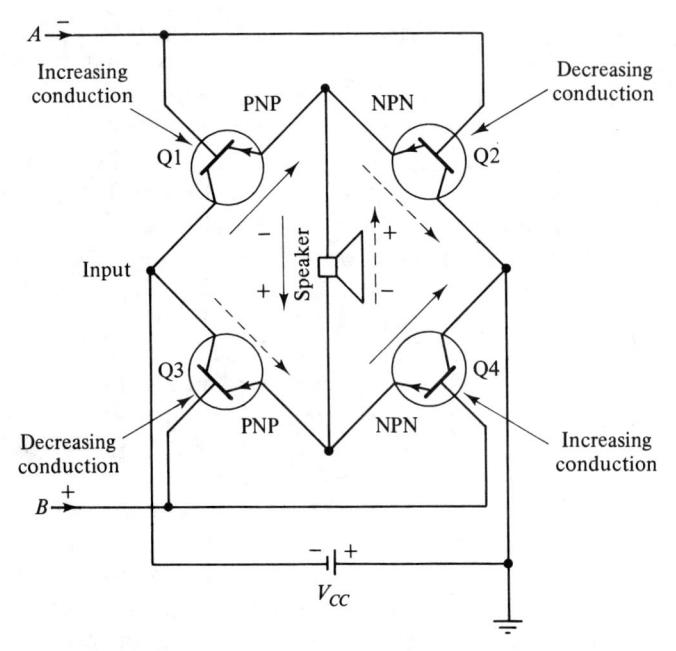

**Figure 5-18.** Complementary-symmetry bridge arrangement.

under idling conditions that the transistors are biased to draw equal currents in class A operation. (This bias circuit is not shown in the diagram.) Under this condition of operation, electrons emerge from the negative battery terminal. One half of the electron current flows through Q1 collector-to-emitter and through Q2 emitter-to-collector, and thence into the positive battery terminal. The other half of the electron current flows through Q3 collector-to-emitter and through Q4 emitter-to-collector, and thence back to the positive battery terminal. Thus the bridge is balanced and there is no current flow through the speaker.

Assume next that an input signal causes point $A$ to become more negative with respect to point $B$. Transistors Q1 and Q4 become more forward biased and draw more collector current. Transistors Q2 and Q3 become less forward biased and draw less collector current. Accordingly, the bridge becomes unbalanced, and the difference in current between Q1 and Q2 flows through the speaker in the direction of the solid-line arrow and through Q4 into the positive battery terminal. A voltage is accordingly developed across the speaker with the indicated polarity. If the input signal causes point $A$ to become positive with respect to point $B$, Q2 and Q3 become forward biased and draw more collector current. Transistors Q1 and Q4 become less forward biased and draw less collector current. The difference between the Q3 and Q4 currents flows through the speaker in the direction of the dashed-line arrow, and develops a voltage drop with the indicated polarity.

### Summary of Power-Amplifier Characteristics

1. The noise factor of an amplifier is defined as the quotient of the signal-to-noise ratio at the output of the amplifier and the signal-to-noise ratio at the input of the amplifier.
2. The noise factor of a transistor increases as its collector voltage is increased. The noise factor, or noise figure, is stated for a given bandwidth and is equal to the ratio of total noise at the output to the noise at the input.
3. A CE amplifier with degeneration develops a comparatively high value of input impedance.
4. Zero-biased class B push–pull amplifiers produce crossover distortion.
5. Crossover distortion can be minimized or eliminated by operating a push–pull amplifier in class AB.
6. Stretching distortion results from application of excessive forward bias in class AB operation.
7. Complementary-symmetry push–pull amplifiers have numerous advantages over related configurations and are widely used. However, transistors must have closely matched characteristics to minimize distortion.
8. Compound-connected or Darlington-connected transistors provide a comparatively high amplification factor.
9. Darlington-connected transistors are frequently used in complementary-symmetry amplifier arrangements to obtain high gain and high power output with minimum circuit complexity.

## 5-7  Examples of Innovative Circuit Design

It is instructive to observe some examples of innovation in audio power amplifier circuit design. As an illustration, class D high-efficiency amplifiers have been devised. The basic principle of class D amplification is shown in Fig. 5-19. Essentially, the input wave form is converted into an amplitude-modulated pulse train that drives the power-output stage. After power amplification of the signal, the pulses are integrated to reconstitute the original input wave form. Higher efficiency is obtained because a power transistor can be driven to considerably higher peak levels when a signal that consists of comparatively narrow pulses is processed.

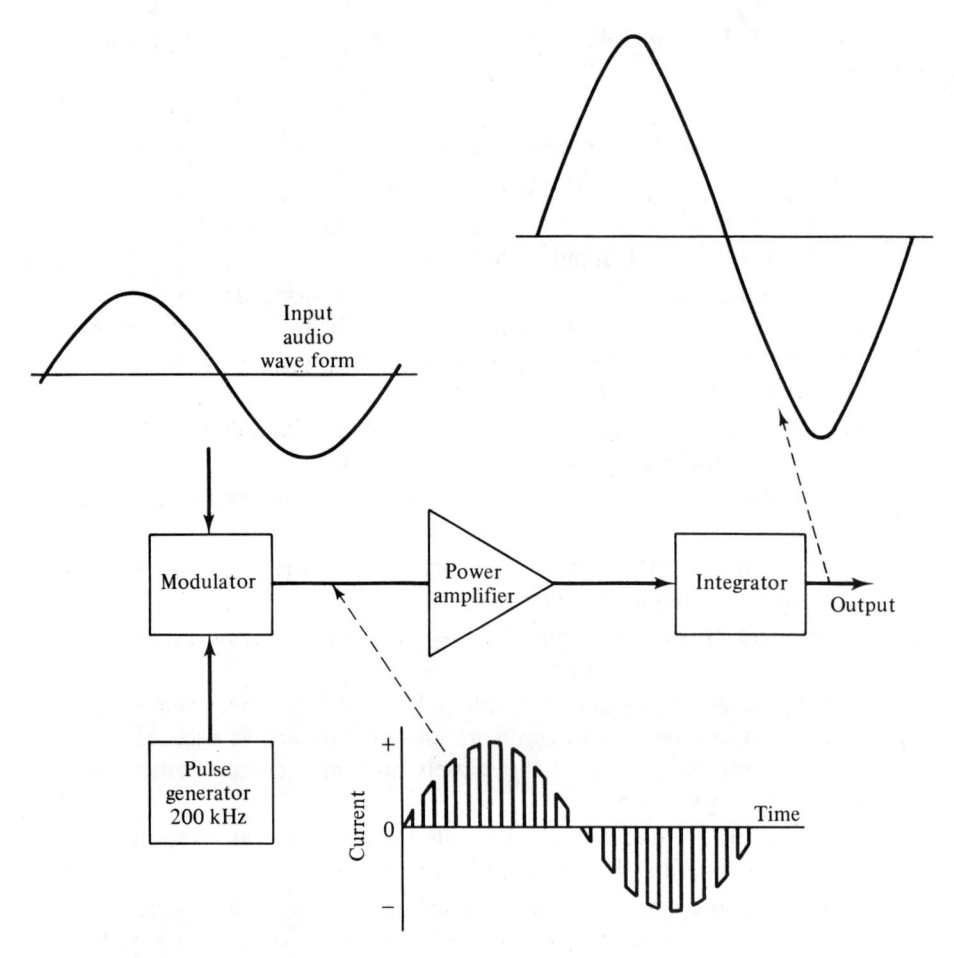

**Figure 5-19.** Principle of a class-D amplifier.

As another example, a pulse-width modulated amplifier also operates at higher efficiency than does a conventional power amplifier. A pulse-width modulated (PWM) wave form is exemplified in Fig. 5-20. This operating technique employs pulses of uniform amplitude and varying width. A narrow pulse corresponds to a low-amplitude input level, whereas a wide pulse corresponds to a high-amplitude input level. A method of PWM wave-form generation is depicted in Fig. 5-21. After power amplification of the PWM waveform, it is integrated to reconstitute the original input waveform. Operating efficiency is comparatively high, because the power transistors can be driven to a relatively high peak level by the pulse waveform.

## 5-7 Examples of Innovative Circuit Design

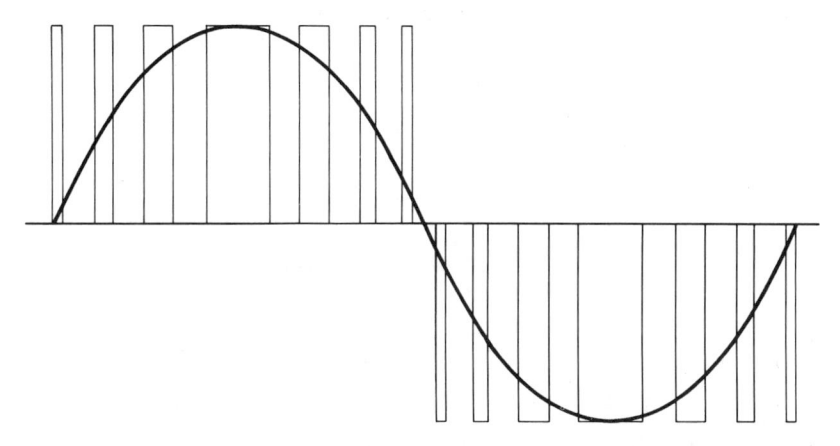

**Figure 5-20.** Example of pulse-width modulated (PWM) wave form.

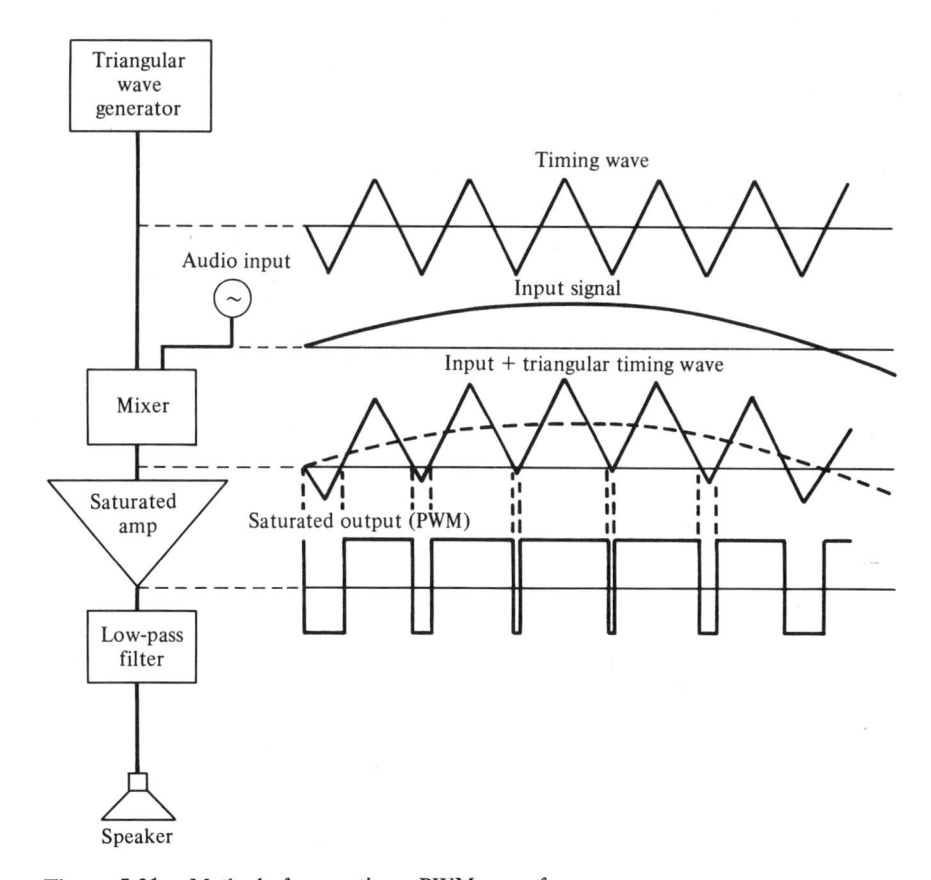

**Figure 5-21.** Method of generating a PWM wave form.

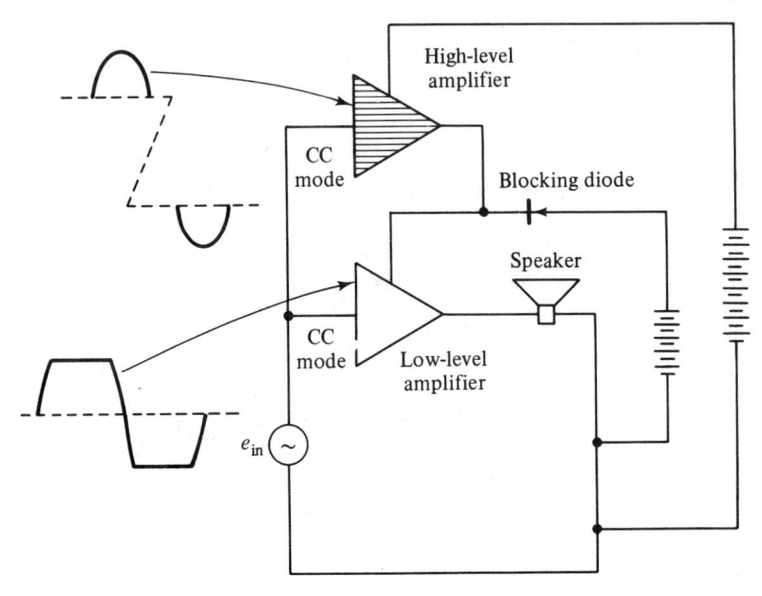

**Figure 5-22.** Basic plan of a class G amplifier.

A new design of high-efficiency high-fidelity amplifier is termed the class G configuration (Fig. 5-22). A skeleton circuit for this form of amplifier is shown in Fig. 5-23. $V_{in}$ denotes the input audio signal; the output is developed across $R_L$ (usually a speaker load). The input signal voltage is applied to the bases of transistors Q1 and Q2. Supply voltage $V_1$ is applied through diode D1 to the emitter of Q1 and to the collector of Q2. In turn, the collector of Q1 is connected to a higher value of supply voltage, $V_{CC}$. When the input signal voltage $V_{in}$ is less than $V_1$, Q1 has a reverse

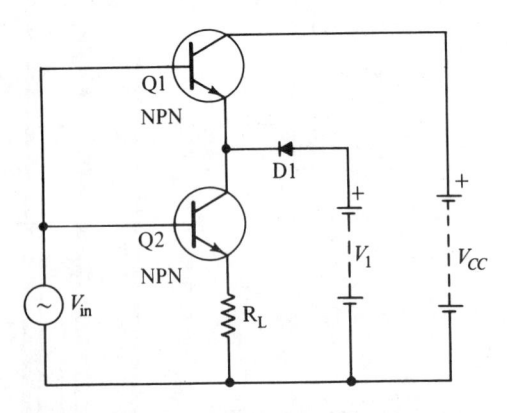

**Figure 5-23.** Skeleton circuit for a class G amplifier.

## 5-7 Examples of Innovative Circuit Design

base–emitter bias, and its collector current is cut off. Current flowing through $R_L$ is obtained from $V_1$ via diode D1.

Next, if the signal voltage increases to a value greater than $V_1$, but less than $V_{CC}$, Q1 becomes forward biased. Its collector current is turned on. Now current flowing through $R_L$ is obtained from $V_{CC}$ via Q1. Diode D1 also has the function of preventing current flow from $V_{CC}$ back into source $V_1$. The operating efficiency of the class G configuration is considerably greater than that of a class B arrangement. However, in its most basic form, a class G amplifier develops an objectionable amount of distortion. This distortion is seen in the waveform depicted in Fig. 5-24. This distortion results from the circumstance that Q1 is not turned on until the amplitude of the input signal voltage exceeds the collector voltage of Q2 by a value equal to the base–emitter voltage of Q1. In turn, Q2 starts to saturate before Q1 begins conduction, and an irregularity is introduced into the output signal waveform.

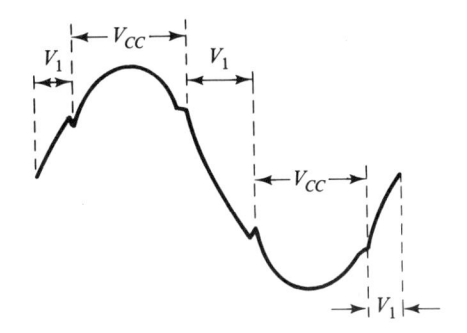

**Figure 5-24.** Distorted output wave form (changeover distortion) developed by a simplified class G amplifier configuration.

This *changeover* distortion can be reduced by adding another diode, D2, as shown in Fig. 5-25. In turn, when $V_1$ has a smaller value than the input signal voltage, the potential between the collector and emitter of Q2 is lower than the saturation level by an amount equal to the threshold value of D2, and Q2 remains unsaturated. Diode D3 also serves an essential function. Inasmuch as a reverse bias voltage is applied between base and emitter of Q1 when the input signal voltage is lower than $V_1$, the base–emitter junction of Q1 must be able to withstand a reverse voltage that is greater than V1. Thus D3 protects the junction against breakdown. A skeleton class G push–pull amplifier configuration is seen in Fig. 5-26. Diodes have been omitted from the base circuits to avoid unnecessary detail. Note that Q3 and Q4 are PNP transistors, whereas Q1 and Q2 are

NPN transistors. In other words, this is a complementary configuration. Its efficiency is considerably greater than that of a conventional complementary arrangement.

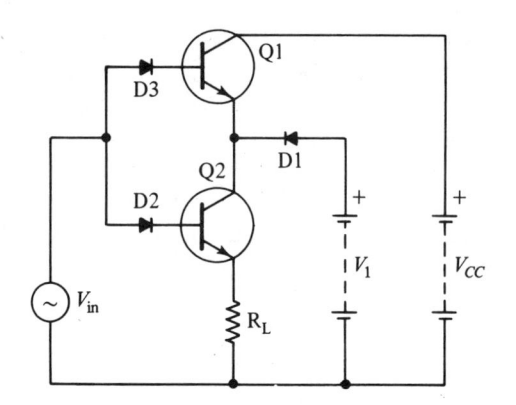

**Figure 5-25.** Distortion is reduced by action of diode D2.

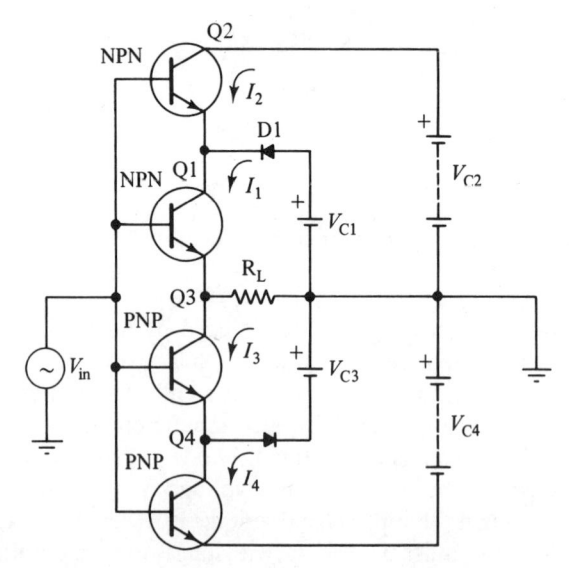

**Figure 5-26.** Skeleton class G push–pull amplifier configuration.

chapter six

# SPEAKER CIRCUITRY

## 6-1 Speaker Interconnections

Speakers in a system may be interconnected in various ways. As an illustration, two or more speakers may be connected in parallel (and phase aiding), as depicted in Fig. 6-1. In this example, each speaker has a rated input impedance of 8 Ω. When two 8-Ω speakers are connected in parallel, their net input impedance becomes 4 Ω. Again, the net input impedance of four 8-Ω speakers connected in parallel becomes 2 Ω. Next consider the net input impedance of a pair of speakers connected in parallel, one of which has a rated input impedance of 8 Ω, whereas the other speaker has a rated input impedance of 4 Ω, as depicted in Fig. 6-1c. These impedance values have a net value of 2.67 Ω. Note carefully that the 4-Ω speaker will draw twice as much audio current as the 8-Ω speaker. In turn, if both of the speakers have the same cone diameters and the same efficiency, the 4-Ω speaker will radiate much more sound energy than the 8-Ω speaker. This would be an undesirable condition in most system arrangements.

With reference to Fig. 6-2, another basic method of speaker interconnection is the series arrangement. This mode of operation results in a net input impedance that is greater than the input impedance of an individual speaker. That is, the system input impedance becomes equal to the sum of the impedances of the individual speakers. Observe that if a low-impedance speaker is connected in series with a high-impedance speaker, the audio voltage drop will be greater across the latter than across the former. Consequently, if both speakers have the same cone diameters and the same efficiency, the high-impedance speaker will radiate more

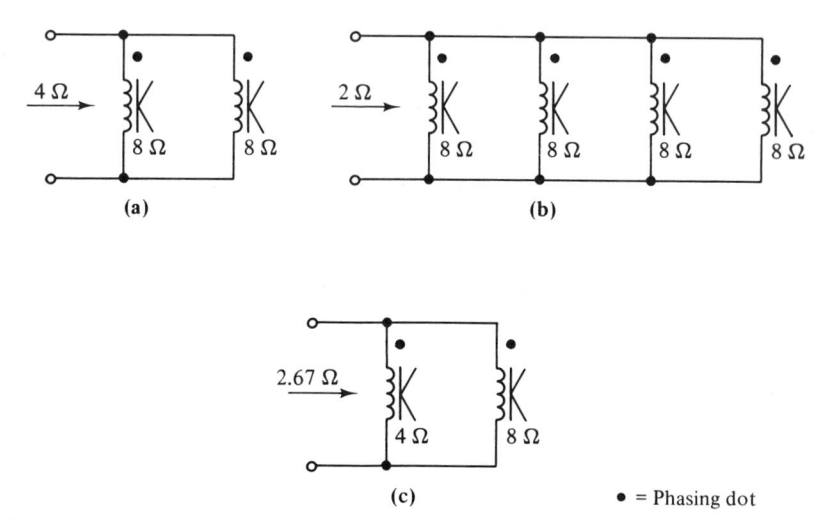

**Figure 6-1.** Examples of parallel-connected speakers: (a) two 8-Ω speakers in parallel; (b) four 8-Ω speakers in parallel; (c) a 4-Ω speaker in parallel with an 8-Ω speaker.

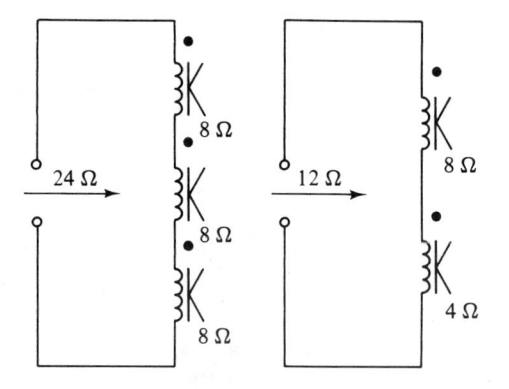

● = Phasing dot

**Figure 6-2.** Examples of series interconnection.

sound energy than the low-impedance speaker. As noted above, this would be an undesirable condition in most system arrangements.

Refer to Fig. 6-3. Still another basic method of speaker interconnection is the series–parallel arrangement. Each speaker in this system has a rated input impedance of 8 Ω. Accordingly, each series string has a net input impedance of 24 Ω, and the two parallel-connected series strings have a net input impedance of 12 Ω. In the event that all the speakers in the system are identical, each speaker will consume the same amount of audio power. On the other hand, if the speakers have different impedances, they

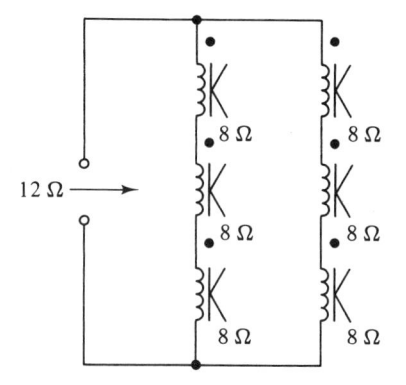

**Figure 6-3.** Series–parallel interconnection arrangement.

will draw different amounts of audio power. The principles outlined above can be applied in this situation to calculate the relative power values drawn by each speaker.

## 6-2  Crossover Circuitry

When a tweeter is connected with a woofer in a speaker system, neither a simple series connection nor a simple parallel connection is practical. In other words, a tweeter cannot withstand high power at low frequencies. Again, a woofer is very inefficient at high frequencies, and largely wastes audio power that is applied. Accordingly, crossover networks are utilized to optimize operating efficiency and to avoid speaker damage. A crossover network is an electrical filter that separates the amplifier output signal into two or more frequency bands for operation of a multispeaker system. The crossover frequency is defined as the frequency at which equal power is delivered to each of the adjacent frequency channels when all channels are terminated in their specified load values.

A very simple crossover arrangement (which is sometimes adequate) is depicted in Fig. 6-4. A basic parallel connection is utilized, but with a series capacitor separating the two speakers. It is evident that at some frequency the reactance (impedance) of the capacitor will be equal to the input impedance of the tweeter. This is the crossover frequency of the network. The value of the crossover frequency is determined by the value of the capacitor. As an illustration, suppose that a 4-$\mu$F capacitor is connected in series with an 8-$\Omega$ tweeter. In this example, the crossover frequency will be approximately 5 kHz. Or, if an 8-$\mu$F capacitor is used, the crossover frequency will be approximately 2.5 kHz. Paper capacitors are utilized in the better-quality crossover networks. However, electrolytic capacitors may also be employed, provided that they are of the nonpolarized type.

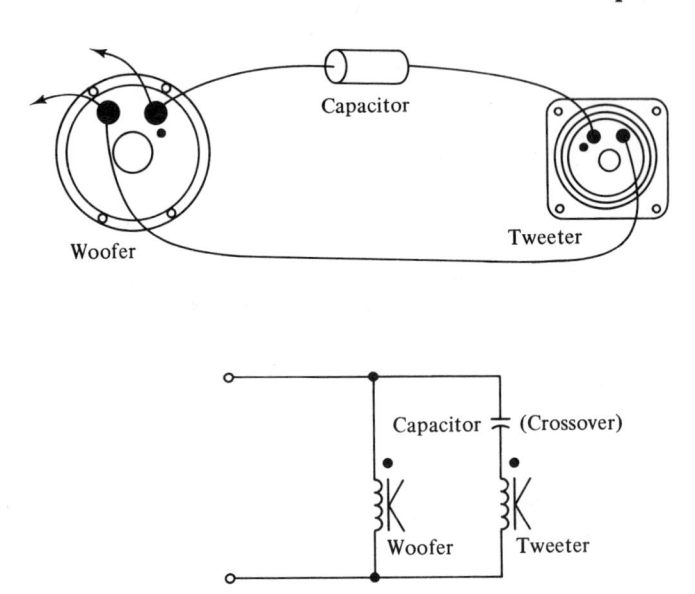

**Figure 6-4.** Simple crossover arrangement.

At the crossover frequency, the audio power delivered to the tweeter is equal to one half of the audio power that is applied to the capacitor–speaker combination. Otherwise stated, the power demand of the tweeter decreases by 3 dB from its high-frequency limiting value to its crossover-frequency value. In turn, the series capacitor prevents low audio-frequency energy from entering the tweeter. In this simple arrangement, high audio-frequency energy is not prevented from flowing into the woofer. Consequently, if the woofer happens to be of a type that can reproduce audio frequencies appreciably above the crossover value, both of the speakers will reproduce midrange audio frequencies, with the result that the midrange audio frequencies may be overemphasized. For this reason, many crossover networks include an inductor connected in series with the woofer, as exemplified in Fig. 6-5.

An inductor has the opposite reactance characteristic with respect to a capacitor. Thus, if the operating frequency increases, the reactance (impedance) of an inductor increases. An inductor with a value of approximately 0.25 millihenry (mH) will have an impedance of 8 Ω at 5 kHz. Again, if the inductance value is increased to 0.5 mH, the crossover frequency will be approximately 2.5 kHz. Consider the net input impedance to a speaker system with an LC crossover network. For example, if 8-Ω speakers are employed in the arrangement of Fig. 6-5, it might be incorrectly concluded that the net impedance of the speaker system would be 4 Ω. However, the fact that the inductor and the capacitor

## 6-2 Crossover Circuitry

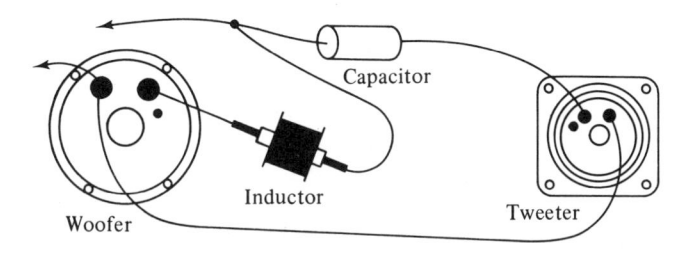

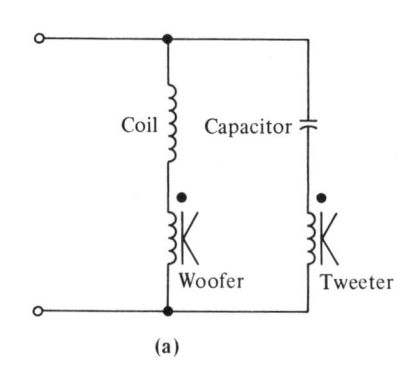

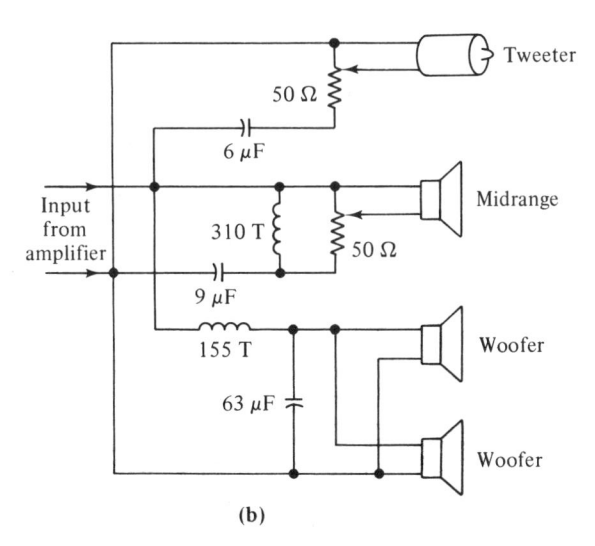

**Figure 6-5.** Examples of crossover networks: (a) woofer–tweeter crossover network; (b) woofer–midrange–tweeter crossover network, with level controls for midrange and tweeter. (*Courtesy, Radio Shack, a Tandy Corporation Company*)

have different reactances (impedances) at various frequencies must be taken into account. It can be shown by ac circuit calculations that the crossover network depicted in Fig. 6-5 provides a practically constant net input impedance to the speaker system; this impedance value is 8 Ω when proper $L$ and $C$ values are utilized.

Refer to Fig. 6-6. As shown by the examples, the net input impedance of the system remains virtually 8 Ω at 10 times and one tenth of the crossover frequency. A speaker has some inductance, in addition to the resistance of its voice coil. This inductive component causes the speaker input impedance to rise somewhat as the operating frequency increases. In turn, the precise impedance values could be somewhat different from those indicated in Fig. 6-6. This is not always the case, however, inasmuch as when an inefficient tweeter is used with a woofer the tendency of the woofer impedance to increase at high frequencies will then be offset by the decreasing input impedance to the tweeter. If a very simple crossover is used, as in Fig. 6-4, the rising input impedance of the woofer at high frequencies operates to prevent the net input impedance of the system from falling below 8 Ω at the high end of the audio band.

Various types of crossover inductors are utilized. A simple form of 0.5-mH inductor is shown in Fig. 6-7. No. 18 wire is employed to minimize $I^2R$ loss. As noted previously, room acoustics can vary considerably

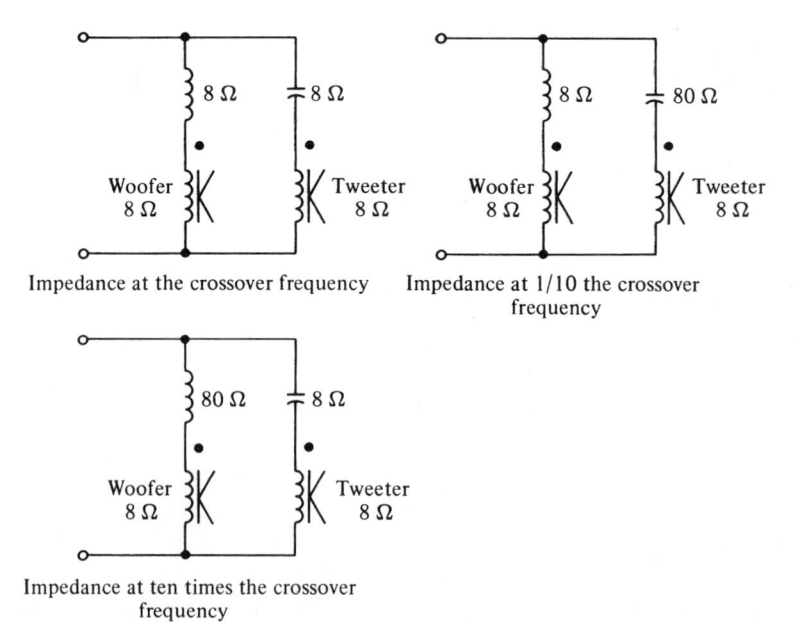

**Figure 6-6.** Examples of system input impedances at three different audio frequencies.

## 6-2 Crossover Circuitry

(a)

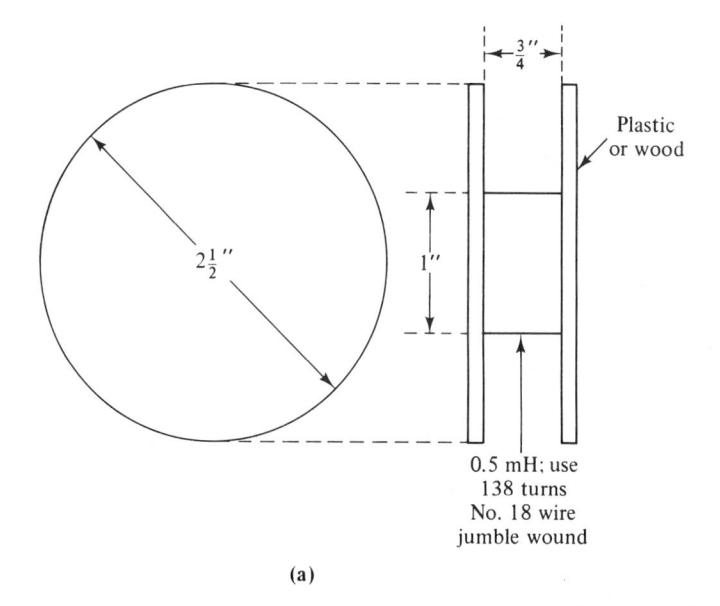

0.5 mH; use
138 turns
No. 18 wire
jumble wound

Note: Use 100 turns for 0.25 mH; use 198 turns for 1 mH

The inductance of a multilayer coil of rectangular cross section can be computed from the formula

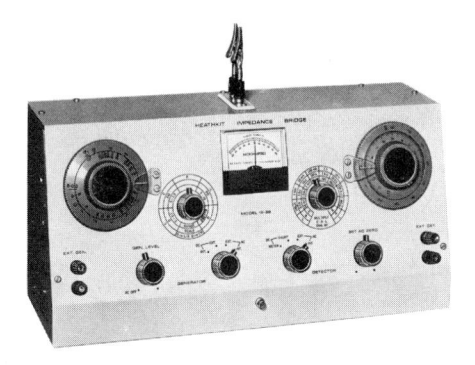

$$L = \frac{0.8(N \times A)^2}{6A + 9B + 10C}$$

where  $L$ = inductance in microhenries
$N$ = number of turns
$A$ = mean radius in inches
$B$ = length of the coil in inches
$C$ = depth of the coil in inches

(b)

**Figure 6-7.** Construction of crossover inductors: (a) winding form and number of turns for typical inductance values; (b) an inductance bridge used to measure inductance values. (*Courtesy, Heath Co.*)

from one location to another. Furthermore, some designs of tweeters are more efficient than others. Consequently, to achieve optimum tonal balance, it is general practice to include a level control in the tweeter branch, as exemplified in Fig. 6-8. Tweeters are usually more efficient than woofers, so that audio current flow to the tweeter must usually be attenuated to some extent. A wire-wound rheostat or potentiometer rated for at least 2 W of power dissipation should be utilized. A total resistance of 50 Ω is

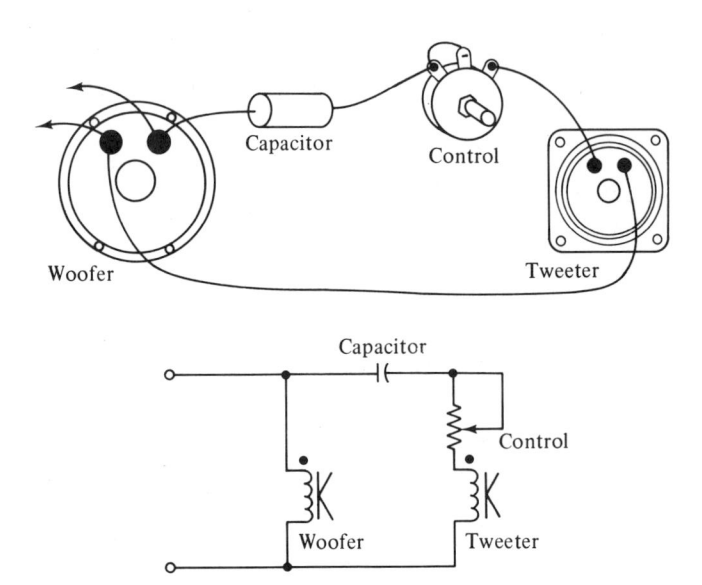

**Figure 6-8.** Level control in a tweeter branch.

suitable, unless a horn type tweeter is employed; in such a case, a resistance range up to 150 Ω is likely to be required. Note that the power rating of a speaker system is equal to the power rating of the woofer plus the power rating of the tweeter. For example, if the woofer has a power rating of 75 W, and the tweeter has a power rating of 25 W, the system will have a power rating of 100 W.

## 6-3  Speaker Phasing

It is essential to connect a pair of speakers in-phase for system operation. Otherwise, the radiation from one speaker will tend to cancel the radiation from the other. This cancellation becomes most apparent in the reproduction of bass tones. For in-phase operation, the speaker cones must move in the same direction at the same time when a voltage is applied at the input of the speaker system. With reference to Fig. 6-9, most speakers have a red dot at one terminal for indicating phase in network circuits. In the case of a parallel connection, the red-dot terminals of the speakers are connected together. In the case of a series connection, the red-dot terminal on one speaker will be connected to the contrary colored terminal on the other speaker (this might be a black dot, for example).

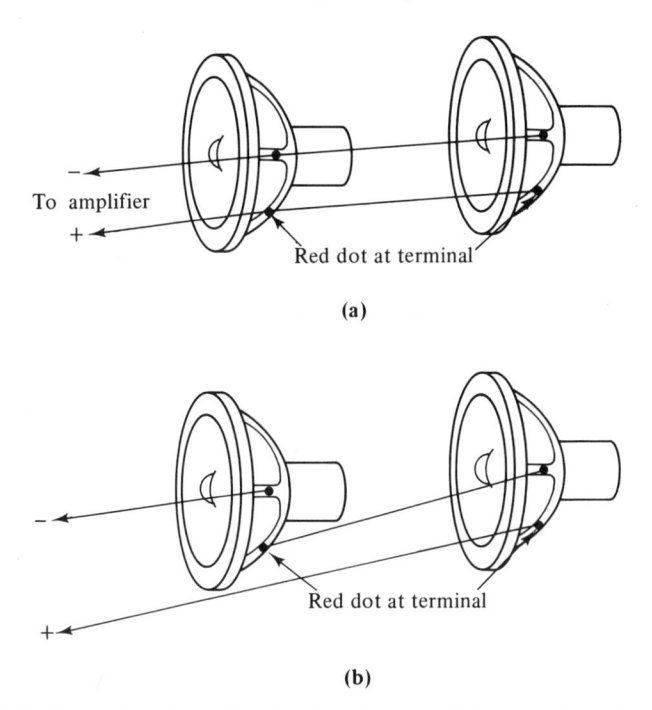

**Figure 6-9.** Examples of speaker phasing: (a) parallel connection; (b) series connection.

## 6-4 Connections for 70.7- and 25-Volt Speaker Systems

Utility sound systems often utilize a 70.7- or 25-V arrangement in order to minimize $I^2R$ losses in the audio lines. The 70.7-V system is preferred for high-power public-address (PA) networks. A 70.7-V matching network for three speakers is shown in Fig. 6-10. A matching transformer is installed to match the particular speaker, or a group of speakers, to the 70.7-V line. This is an example of a *constant-voltage system,* because the line voltage is comparatively unaffected by switching individual speakers on and off. Network calculations are made as follows:

1. Determine the power rating of each speaker.
2. Add the power values to find the total power demand; use a 70.7-V amplifier with a rated power output at least equal to this demand.
3. Select a 70.7-V matching transformer for each speaker (or for each group of speakers) with appropriate primary wattage ratings.

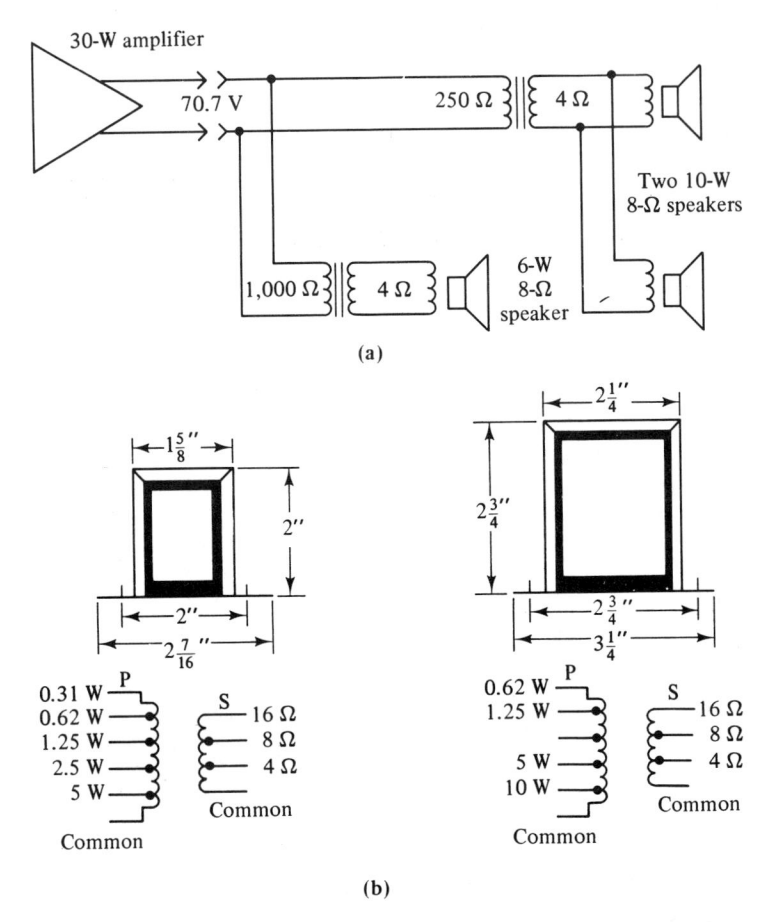

**Figure 6-10.** Example of a 70.7-V speaker-matching network: (a) configuration; (b) typical 70.7-V transformer taps and dimensions.

4. Connect the primary terminals of each transformer across the 70.7-V line from the amplifier output. A primary mismatch up to ±25 percent is tolerable.
5. Connect the secondary terminals of each transformer to its speaker (or group of speakers), observing the matching ohms tap.
6. In the event that the matching transformers are rated in impedance values, the primary wattage for a transformer may be calculated as follows:

$$Z_p = \frac{(70.7)^2}{P}$$

## 6-4 Connections for 70.7- and 25-Volt Speaker Systems

where $Z_p$ is the rated primary impedance and $P$ is the wattage rating of the speaker.

It follows that basic power and impedance ratings are as follows:

1 watt corresponds to 5,000 Ω $Z_p$

2 watts correspond to 2,500 Ω $Z_p$

5 watts correspond to 1,000 Ω $Z_p$

10 watts correspond to 500 Ω $Z_p$

With reference to Fig. 6-10 the 6-W speaker has a 4-Ω voice coil, and the paralleled 10-W speakers have 8-Ω voice coils. Accordingly, the total power demand is 26 W and the amplifier would be rated for approximately 30-W output. In accordance with the foregoing equation, the primary impedance for the 6-W transformer would be 833 Ω; a 1,000-Ω impedance could be utilized. For the 20-W speaker combination, the primary impedance will be 250 Ω.

chapter seven

# BASIC TELEPHONE CIRCUITRY

## 7-1 General Considerations

Almost everyone is familiar with the sound-powered telephone circuit shown in Fig. 7-1a. A telephone receiver is connected at each end of a line to serve alternately as a receiver or as a transmitter (microphone). A more efficient arrangement employs a separate carbon transmitter, as depicted in Fig. 7-1b. A carbon transmitter is a simple form of audio amplifier. In turn, voice currents can be carried over lines to greater distances. Improved operation is provided by the *induction coil I* in the diagram. An induction coil is essentially an impedance-matching transformer that matches the low impedance of the carbon microphone to the higher impedance of the line. An elementary telephone circuit with ringer facilities is depicted in Fig. 7-1c. A 20-Hz ac source is provided for energizing the two ringers (bells) on the line. Because of the shunt signal paths in the configuration, efficiency is less than could be realized. However, this is a *tradeoff* that is accepted by the designer in order to minimize the switching circuitry that is utilized.

An undesirable feature of the simple arrangement depicted in Fig. 7-1 is the presence of *sidetone* in the receiver at the transmitting location. Sidetone is defined as the reproduction, in a telephone receiver, of the sounds produced by the transmitter of the same telephone set; that is, a loud reproduction of one's own voice in the receiver of a telephone set when speaking into the mouthpiece. Therefore, telephone circuit designers employ antisidetone circuitry with a hybrid coil, as shown in Fig. 7-2. A hybrid coil, also called a bridge transformer, has effectively three windings. It is connected into the four branches of the circuit so that voice

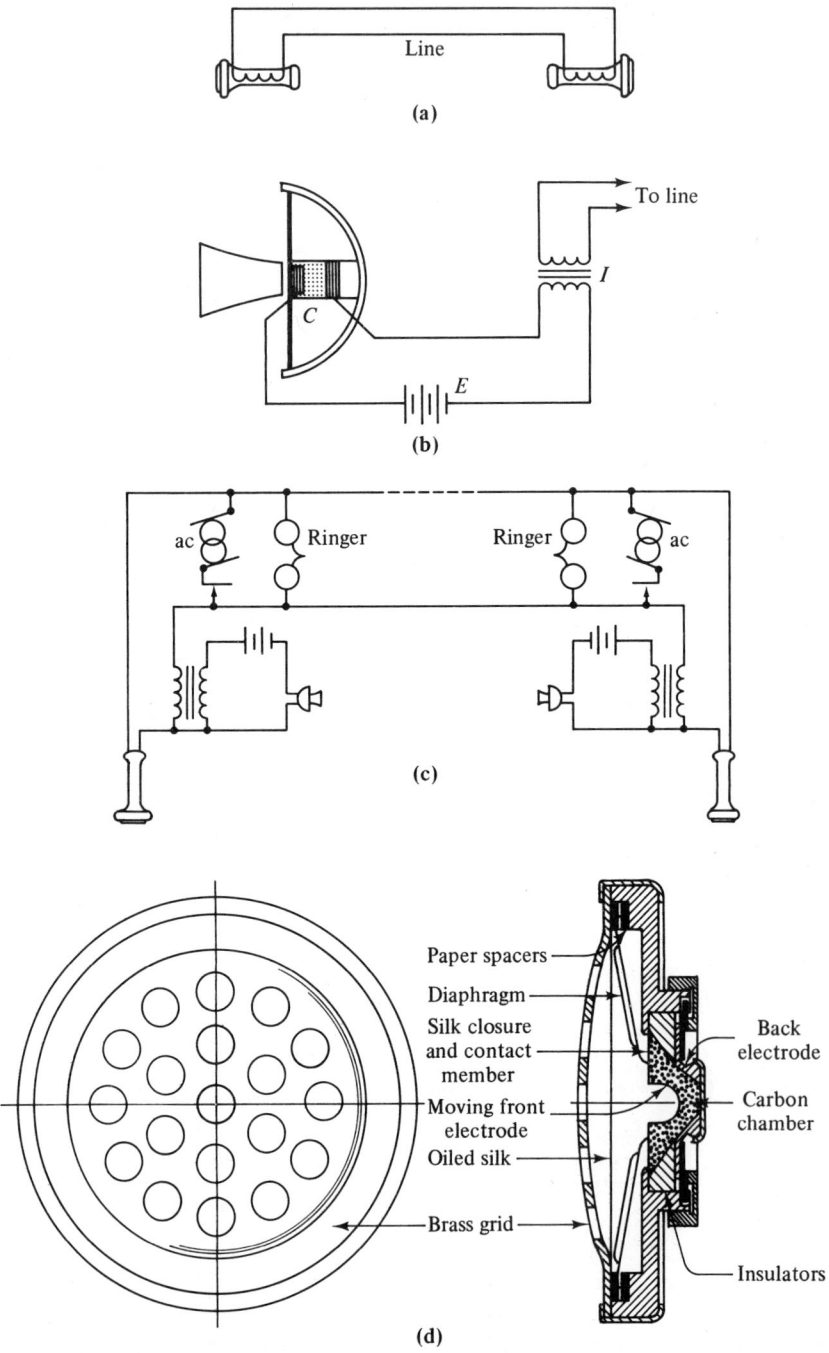

**Figure 7-1.** Fundamental telephone arrangements: (a) sound-powered telephone circuit; (b) carbon transmitter (microphone) circuit; (c) elementary telephone circuit with ringer facilities; (d) design of modern carbon transmitter.

## 7-1 General Considerations

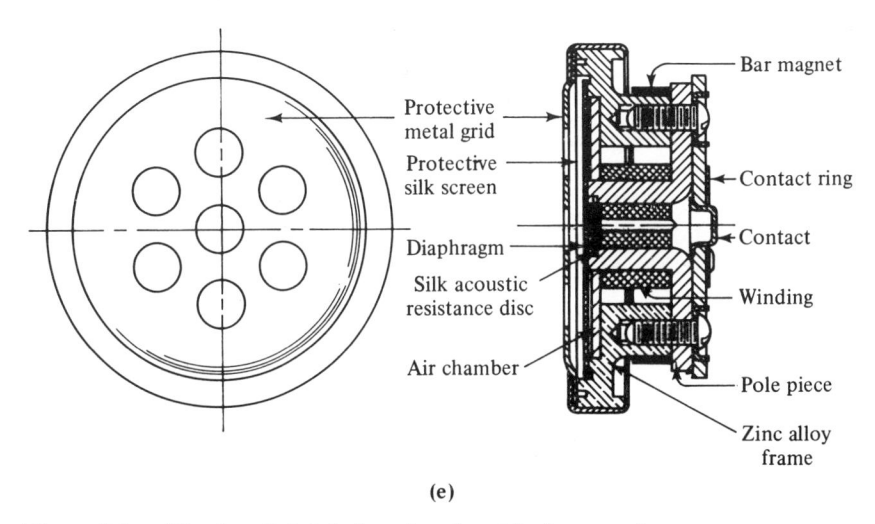

Labels (figure e): Bar magnet, Protective metal grid, Protective silk screen, Contact ring, Diaphragm, Contact, Silk acoustic resistance disc, Winding, Air chamber, Pole piece, Zinc alloy frame

(e)

**Figure 7-1.** (*Continued*) (e) design of modern telephone receiver.

currents from the transmitter flow into the line, but are canceled in the receiver of the same telephone set. This advantage is obtained at the expense of a tradeoff in efficiency, because half of the voice current from the transmitter is dissipated in the hybrid-coil configuration.

With reference to Fig. 7-2a, a dc Wheatstone bridge is depicted. The four resistors, R1 through R4, called the arms of the bridge, are connected to a battery. A galvanometer, a current-indicating instrument, is connected to the junction points $A$ and $B$. Current from the battery is supplied to resistor branches R1–R2 and R3–R4; consequently, there is a potential difference across each of the resistors. When, in addition, there is a potential difference between points $A$ and $B$, current passes through the galvanometer and causes its pointer to deflect. The bridge is said to be balanced when there is no potential difference between points $A$ and $B$, and the galvanometer does not deflect. This occurs when the ratio of R1 to R2 equals the ratio of R3 to R4. Thus, when the bridge is balanced, there is no current through the galvanometer.

In Fig. 7-2b, the resistors are replaced by impedances, which, in addition to resistance, have inductive or capacitive reactance. The battery is replaced by an ac sine-wave generator, and the galvanometer is replaced by a telephone receiver. This arrangement constitutes an ac bridge circuit, which, when balanced, passes no current through the receiver. The ac voltage supplied by the generator is of fixed frequency. When the bridge is not balanced, there is a potential difference between points $A$ and $B$, and this produces an alternating current through the receiver and causes it to generate sound waves. When the ratio of the impedance Z1 to Z2 equals that of Z3 to Z4, however, the bridge is balanced, and there is neither current in nor sound from the receiver.

In place of the generator, a transmitter and a battery may be used as the source of ac voltage, as shown in Fig. 7-2c. Also, in place of impedance Z4, the impedance of a telephone line and connected telephone set, Z(line), may be utilized. If Z(line) equals Z4, the bridge is still balanced, because the ratio of impedance Z1 to Z2 now equals that of Z3 to Z(line). When a voice-current voltage is generated by the transmitter, there is no current through the receiver and it generates no sound, but there is a voice current in each of the impedances of the bridge circuit, one of which is the telephone line and the connected telephone set Z(line). This voice current is reproduced by the receiver of the telephone set, but not by the telephone receiver of the bridge circuit. On the other hand, voice current from the transmitter of the connected telephone set produces a response in the bridge receiver. This response occurs because the bridge is not balanced for a voltage applied between terminals L1 and L2.

The bridge arrangement in Fig. 7-2c can be used as the circuit of a telephone set that eliminates sidetone, provided that the condition of balance is satisfied at all voice-current frequencies. In practice, it is impossible to obtain an exact balance over the entire range of voice frequencies, because the impedances of the components in the circuit vary with frequency; but, for any given telephone line, the balance can be so adjusted that the sidetone is small. Since the impedance of the transmission line is considered to be one of the components of the antisidetone circuit, the efficiency of the circuit depends in part on the length of the line that connects the two telephone sets. When transmitting with this circuit, part of the energy of the voice current is wasted in the impedance arms Z1 and Z2, as well as in Z3 and Z(line). However, by substituting windings N1 and N2 of a transformer or induction coil, as shown in Fig. 7-2d, the loss of energy can be reduced. If N1 and N2 were the windings of an ideal transformer (if they had negligible resistance), no energy would be used in impedances Z3 and Z(line).

The induction coil in Fig. 7-2d can be replaced by one that has a separate primary winding, N, as in Fig. 7-2e. This does not change the bridge action of the circuit, but it permits the voltage of the voice current that is generated by the transmitter to be stepped up or stepped down in any desired ratio by choice of the proper number of primary turns. Thus a transmitter of any resistance can be used efficiently. In an ideal telephone set, the element Z3, which may consist of a noninductive resistance, is combined physically with the resistance of winding N1. The circuit with the type of induction coil called an *autotransformer* is shown in Fig. 7-2f. An autotransformer is a transformer with a primary and secondary that are parts of the same winding. An ac voltage applied across a portion of the turns of the autotransformer induces an ac voltage in the

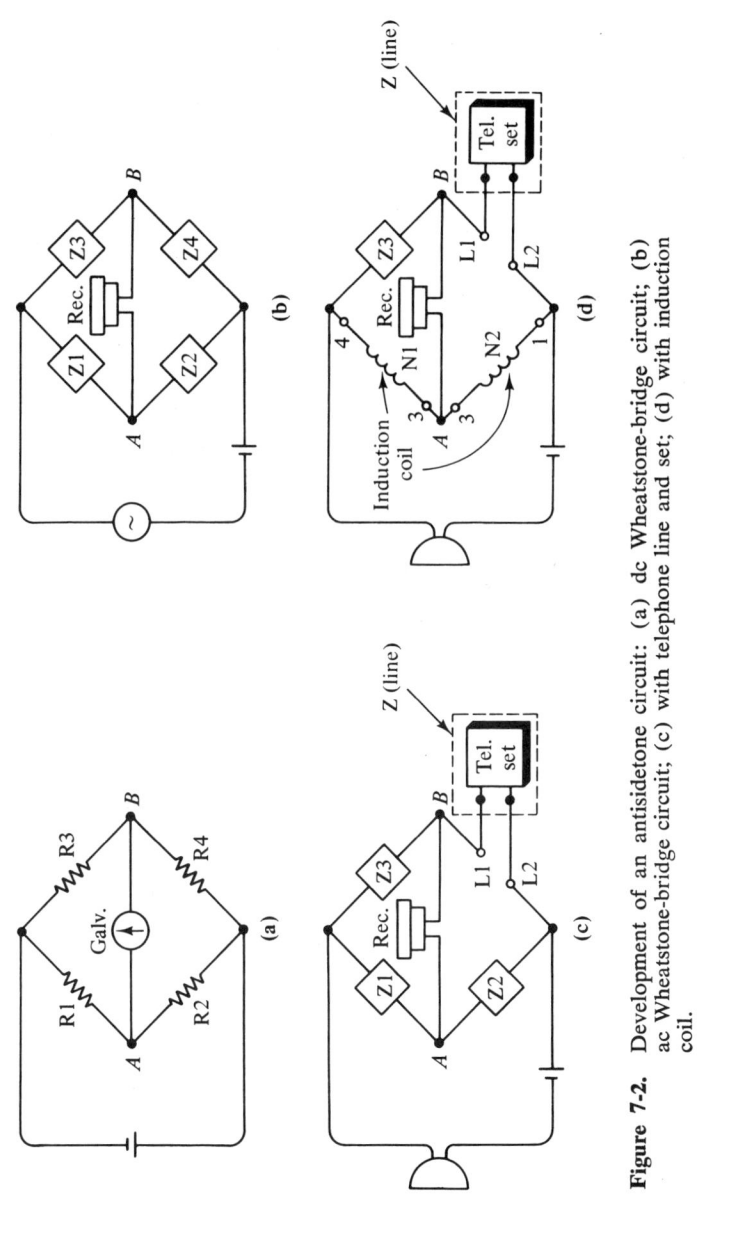

**Figure 7-2.** Development of an antisidetone circuit: (a) dc Wheatstone-bridge circuit; (b) ac Wheatstone-bridge circuit; (c) with telephone line and set; (d) with induction coil.

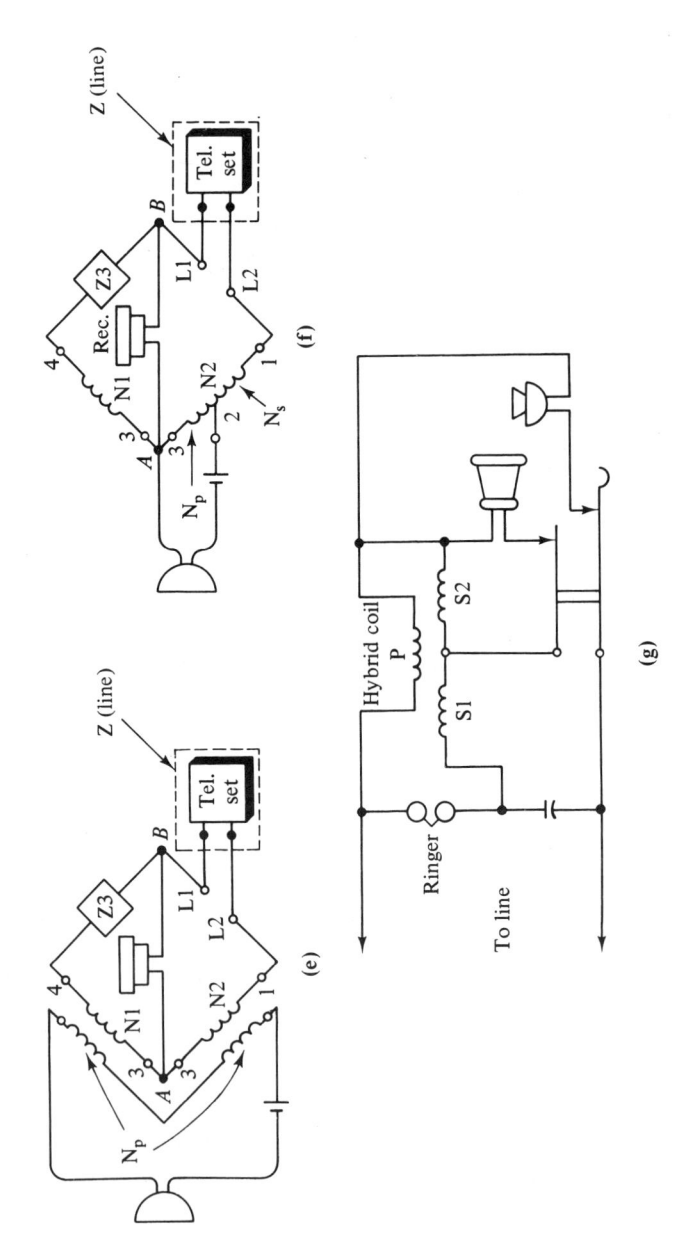

**Figure 7-2.** (*Continued*) (e) with transformer; (f) with autotransformer-type induction coil; (g) station circuit.

full winding, because the autotransformer is so designed that the varying magnetic flux in the turns to which the voltage is applied links all the other turns.

As in the case of a standard transformer, the primary winding is the one to which the voltage is applied, and the secondary is the one in which the voltage is induced. N1 and N2 in Fig. 7-2f are parts of a single winding, with intermediate connections at points 2 and 3. The voltage of the voice current generated by the transmitter is applied across points 2–3, a part of winding N2. This part of the winding, N2, is therefore the primary winding, Np, of the autotransformer. The varying voice current in Np induces a voltage in N1–N2. This induced voltage is greater than that applied to the primary winding, Np, because winding N1–N2 has more turns than winding Np. The use of an induction coil of the autotransformer type produces a stepup in the voltage of the voice current, which appears at terminals 1–4 of secondary winding N1–N2. In this circuit, since the transmitter is connected across the same number of turns (Np) as in the circuit shown in Fig. 7-2e, the action is essentially the same. However, use of an autotransformer results in a slightly more efficient circuit. The station-circuit version of this arrangement is exemplified in Fig. 7-2g.

## 7-2  Phantom Circuits

When two pairs of wires of the proper type are available, a third transmission path may be obtained in a system by using one pair of wires for one side of the third circuit and the second pair for the other side. This basic arrangement is shown in Fig. 7-3. Repeating coils are used in a bridge configuration. For satisfactory phantom operation on open-wire lines, it is necessary that the two wires of a pair have approximately equal resistances and that the lines be suitably *transposed* to prevent objectionable *crosstalk* among the three constituent circuits of each phantom group and between nearby phantom groups in the system.

In cable circuits, crosstalk considerations require the use of *quadded* conductors, suitably spliced at suitable intervals to minimize side-to-side and phantom-to-side crosstalk. A quadded cable is defined as a cable in which some or all the conductors are in the form of quads, consisting of four separately insulated conductors twisted together. A *spiral-four quad* consists of four wires laid together and twisted as a group, the diagonally opposite wires being used as a pair. Domestic lead-covered quadded cables are of multiple-twin design, consisting of two twisted pairs, with the pairs twisted together. The phantom-deriving repeating coils must be well balanced. Phantom circuits tend to be noisier than their side circuits, or non-phantomed circuits, particularly if the lines or equipment are not maintained in good electrical balance.

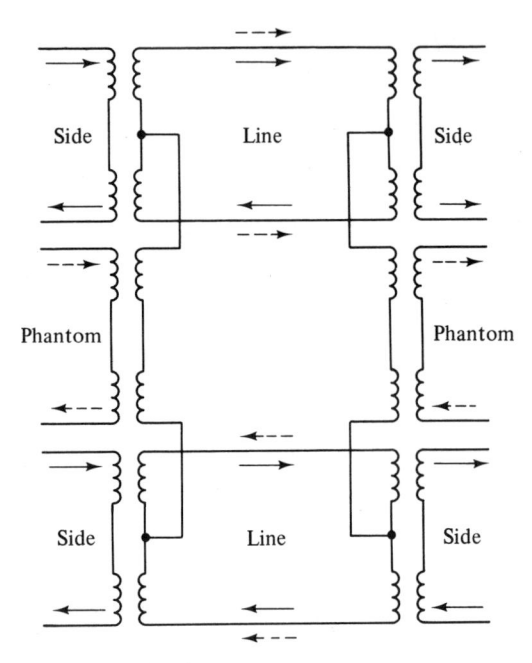

**Figure 7-3.** Basic phantom circuit.

## 7-3 Loading Principles

Cables and lead-covered paper-insulated cables may be operated with or without *loading*. Loading is the addition of series inductance at regular intervals along a line. This added inductance increases the impedance of the circuit and decreases the series losses due to conductor resistance. Loading increases any shunt losses caused by leakage and also causes the line to have *a cutoff frequency* above which the line loss becomes very high. The loss is actually increased by loading at frequencies above about 90 percent of the cutoff frequency. The approximate cutoff frequency and the nominal impedance of a loaded line are given by the following equations:

$$f_0 = \frac{1}{\pi\sqrt{LC}}$$

$$Z = \sqrt{\frac{L}{C}}$$

where

$f_C$ = cutoff frequency in hertz

$Z$ = nominal impedance (resistance) in ohms

$L$ = inductance of a loading coil in henrys

$C$ = capacitance of a loading section in farads

The usual method of designing loading systems is by first indicating the spacing of the loading coils in feet, then the inductance of the side circuit in millihenries, and last the inductance of the phantom circuit (if there is one) in millihenries. For example, 6,000–88–50 represents a loading system in which the loading coils are spaced 6,000 ft apart, the inductance of the side-circuit loading coils is 88 mH, and the phantom circuit employs loading coils with 50-mH inductance. Again, 6,000–88 denotes a nonphantomed circuit loaded with 88-mH loading coils spaced 6,000 ft apart. A speech-frequency circuit could consist of 88-mH loading coils spaced at 1-mile intervals. However, a carrier-frequency circuit requires a higher cutoff frequency; it could consist of 6-mH loading coils spaced at quarter-mile intervals. The extent of the system designer's problem becomes apparent when it is recognized that the capacitance of a cable circuit between San Francisco and New York is approximately 180 $\mu$F.

## 7-4 Repeaters

The range of voice-frequency circuits can be extended by the use of line amplifiers called *repeaters*. Three basic types are termed the 21-type, the 22-type, and the four-wire repeater. A 21-type repeater, depicted in Fig. 7-4, has a circuit arrangement that requires no balancing networks, and stability (freedom from singing) is realized by the balance between the

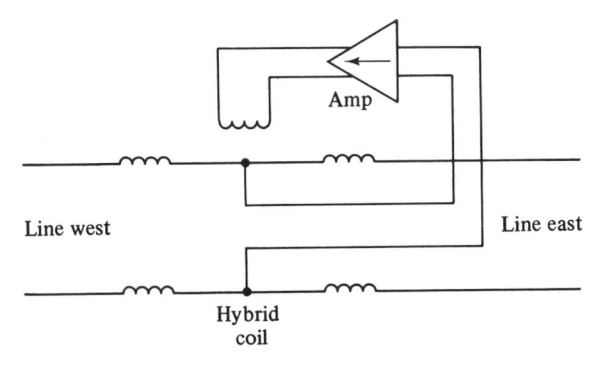

**Figure 7-4.** Arrangement of a 21-type repeater.

lines on the two sides of the repeater. Since freedom from self-oscillation (singing) depends upon equality of impedances on the two sides of the repeater, this configuration is not suitable for use on circuits that are made up from more than one kind of facility. Thus the best location for a 21-type repeater is at the midpoint of a line or circuit. It is impossible to use a 21-type repeater on a loaded circuit, but the usable gain is subject to wide variations, and in some cases it may be very small. Type-21 repeaters can be operated in tandem, but in general there is little transmission advantage in this expedient.

A 22-type repeater, depicted in Fig. 7-5, employs a circuit arrangement with two balancing networks. Stability is obtained by the balance between the impedance of each network and its associated line. This type of repeater may be used at a circuit terminal or at intermediate points, and will operate satisfactorily in tandem with other 22-type repeaters. A 22-type repeater can be used on any type of stabilized line for which suitable balancing networks are available. These repeaters may also be used on nonstabilized lines, but at considerably reduced gain. The maximum usable gain (MUG) of a 22-type repeater is limited to a value that provides adequate margin against singing or near singing. In some systems, crosstalk or echoes, rather than singing, may limit the usable gain.

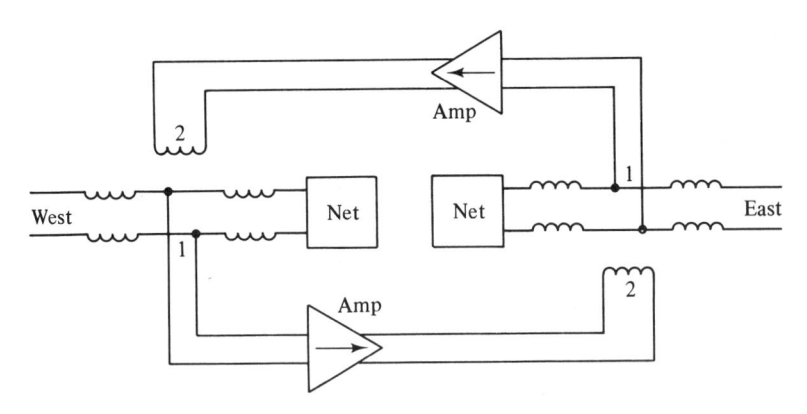

**Figure 7-5.** Arrangement of a 22-type repeater.

A four-wire repeater (Fig. 7-6) consists of two one-way amplifiers that operate in opposite directions. Each amplifier processes signals in one direction only. Four-wire repeaters can be used at circuit terminals and in tandem with other four-wire repeaters. It is an inherently stable arrangement and can provide substantial gain on a line that has irregular impedance. Usable gain is generally limited by crosstalk, noise, or transmission variations. No balance requirements are involved. On the other

## 7-5 Return Loss

hand, a 21-type or a 22-type repeater requires careful balance. Refer to Fig. 7-5. If the two amplifiers were merely connected together, an oscillator configuration would result, and the arrangement would sing or howl. This difficulty is avoided by the use of a bridge circuit arrangement that effectively isolates the two amplifiers.

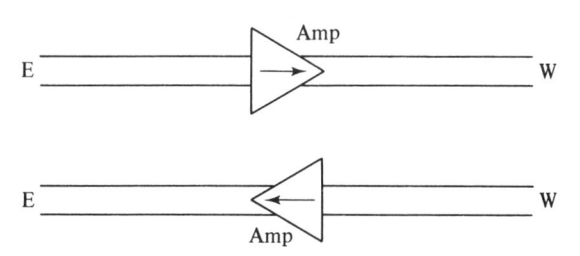

**Figure 7-6.** Four-wire repeater arrangement.

The action of a hybrid coil is similar to that of a Wheatstone bridge. When a voltage is applied across a particular pair of terminals of the bridge, no current will flow through an impedance connected across the other two terminals of the bridge, if balance conditions regarding the ratios of the impedances of the four arms of the bridge are observed. Thus, if a voltage is impressed on terminal 2 in Fig. 7-5, no current will flow in terminal 1 if the impedances of the line and network are identical. The greater the difference between the impedances of line and network, the greater will be the amount of current that flows in terminal 1 and the greater will be the tendency of the repeater to sing. Since perfection cannot be attained even in the most carefully designed telephone circuits, there is always some transmission across the hybrid coil. Consequently, the amount of amplification that can be provided by a repeater without singing is limited. The sum of the gains of the two amplifiers must always be greater than the sum of the losses across the two hybrid coils in order to prevent singing. In practice, *to allow for variations in the circuit, the designer specifies a margin between total loss and total gain.*

### 7-5 Return Loss

The *return loss* between two impedances is a measure of the similarity between the impedances. These might be the line and network impedances or the impedances of two types of line. It is expressed in decibels, and is equal to 20 times the logarithm of the reciprocal of the numerical value of the reflection coefficient, that is, return loss:

$$R = 20 \log \left[ \frac{Z_1 + Z_2}{Z_1 - Z_2} \right]$$

The loss across a conventional hybrid coil is about 6 dB greater than the return loss. Return loss is defined as the difference between the power incident upon, and the power reflected from, a discontinuity in a transmission system, or as the ratio in decibels of these power values. Since speech transmission through a repeater must be practically uniform over the band of frequencies that is employed (approximately 200 to 2,800 Hz), it is evident that transmission loss across a hybrid coil must be sufficiently great at all frequencies in this band to prevent singing at any one frequency. *System designers generally utilize filters to prevent singing at frequencies outside the speech-transmission band.*

It is evident that good balance cannot ordinarily be obtained by matching the line impedance at a single frequency by means of a simple network consisting of a resistor and a capacitor of an inductor. Excellent balance between line and network could, of course, be obtained by duplicating in the network each element of the line. However, this is an impractical solution. Therefore, fixed networks are utilized by the designer, or variable networks are provided that can be adjusted in the field to approximately match the characteristic impedance of the types of lines in general use. These networks are furnished as a part of the repeater.

To obtain good balance between a line and a network, the line must be reasonably uniform. Where apparatus is placed between the hybrid coil and the line, duplicate apparatus is usually placed between the hybrid coil and the network. A line irregularity that is distant from the repeater is less important than one that is close to the repeater. If a repeater section is composed of two dissimilar lengths of line in tandem, L1 and L2, with characteristic impedances Z1 and Z2, and if the repeater connected to L1 matches Z1, and the repeater connected to L2 matches Z2, then the return loss at the junction of L1 and L2 is $R$, as explained above, but the return loss at the repeater connected to L1 is equal to $R + 2A$, where $A$ is the attenuation of L1. This is apparent by noting that a current starting at the repeater would traverse L1, be partially reflected at the junction of L1 and L2, and traverse L1 again before the reflected voltage reaches the repeater.

## glossary

# ACOUSTIC, AUDIO-FREQUENCY, AND SOUND TERMS

**A-B test.**  comparison of sound from two sources, such as comparing original program to tape as it is being recorded by switching rapidly back and forth between them.

**Accompaniment.**  also called *lower* or *great*. The lower manual of an organ, which provides the musical harmony to the solo or melody.

**Acetate backing.**  a standard plastic base for magnetic recording tape.

**Acoustic.**  pertaining to sound or to the science of sound.

**Acoustic absorption loss.**  the energy lost by conversion into heat or other forms when sound passes through or is reflected by a medium.

**Acoustic absorptivity.**  the ratio of sound energy absorbed by a surface to the sound energy arriving at the surface. Equal to 1 minus the reflectivity of the surface.

**Acoustic attenuation constant.**  the real part of the acoustic propagation constant; neper per section, or unit distance.

**Acoustic capacitance.**  in a sound medium, a measure of volume displacement per dyne per square centimeter. The unit is centimeter to the fifth power per dyne.

**Acoustic clarifier.**  a system of cones loosely attached to the baffle of a speaker and designed to vibrate and absorb energy during sudden loud sounds, thereby suppressing them.

**Acoustic compliance.**  the measure of volume displacement of a sound medium when subjected to sound waves. Also, that type of acoustic reactance which corresponds to capacitive reactance in an electrical circuit.

**Acoustic delay line.**  a device that retards one or more signal vibrations by causing them to pass through a solid (or liquid).

**Acoustic dispersion.**  the change in speed of sound with frequency.

**Acoustic elasticity.**  the compressibility of the air in a speaker enclosure as the cone moves backward. Also, the compressibility of any material through which sound is passed.

**Acoustic-electric transducer.** a device designed to transform sound energy into electrical energy, and vice versa.

**Acoustic feedback.** also called acoustic regeneration. The mechanical coupling of a portion of the sound waves from the output of an audio-amplifying system to a preceding part or input circuit (such as a microphone) in the system. When excessive, acoustic feedback produces a howling sound in the speaker.

**Acoustic filter.** a sound-absorbing device that selectively suppresses certain audio frequencies while allowing others to pass.

**Acoustic frequency response.** the voltage-attenuation frequency measured into a resistive load, producing a bandwidth approaching sufficiently close to the maximum.

**Acoustic generator.** a transducer such as a speaker, which converts electrical or other forms of energy into sound.

**Acoustic horn.** also called a horn. A tube of varying cross section having different terminal areas, which change the acoustic impedance to control the directivity of the sound pattern.

**Acoustic impedance.** total opposition of a medium to sound waves. Equal to the force per unit area on the surface of the medium, divided by the flux (volume velocity or linear velocity multiplied by area) through that surface. Expressed in ohms and equal to the mechanical impedance divided by the square of the surface area. One unit of acoustic impedance is equal to a volume velocity of 1 $cm^3$ per s produced by a pressure of 1 $\mu$bar. Acoustic impedance contains both acoustic resistance and acoustic reactance.

**Acoustic inertance.** a type of acoustic reactance that corresponds to inductive reactance in an electrical circuit. (The resistance to movement or reactance offered by the sound medium because of the inertia of the effective mass of the medium.) Measured in acoustic ohms.

**Acoustic intensity.** the limit approached by the quotient of acoustic power being transmitted at a given time through a given area divided by the area as the area approaches zero.

**Acoustic labyrinth.** a special speaker enclosure having partitions and passages to prevent cavity resonance and to reinforce bass response.

**Acoustic lens.** an array of obstacles that refracts sound waves in the same way that an optical lens refracts light waves. The dimensions of these obstacles are small compared to the wavelengths of the sound being focused. Also, a device that produces convergence or divergence of moving sound waves. When used with a speaker enclosure, an acoustic lens widens the beam of the higher-frequency sound waves.

**Acoustic line.** mechanical equivalent of an electrical transmission line. Baffles, labyrinths, or resonators are placed at the rear of a speaker enclosure to assist in reproduction of very low audio frequencies.

**Acoustic mode.** a mode of crystal-lattice vibration that does not produce an oscillating dipole.

**Acoustic ohm.** the unit of acoustic resistance, reactance, or impedance. One acoustic ohm is present when a sound pressure of 1 dyne per $cm^2$ produces a volume velocity of 1 $cm^3$ per s.

**Acoustic phase constant.** the imaginary part of the acoustic propagation constant. The commonly used unit is the radian per second per second or unit distance.

**Acoustic pickup.** in nonelectrical phonographs, the method of reproducing a recording by linking the needle directly to a flexible diaphragm.

**Acoustic radiator.** in an electroacoustic transducer, the part that initiates the radiation of sound vibration. A speaker cone or an earphone diaphragm are examples.

**Acoustic reactance.** that part of the acoustic impedance due to the effective mass of the medium, that is, to the inertia and elasticity of the medium through which the sound travels. The imaginary component of acoustic impedance, expressed in acoustic ohms.

**Acoustic reflectivity.** the ratio of the rate of flow of sound energy reflected from the surface on the side of incidence to the incident rate of flow.

**Acoustic refraction.** a bending of sound waves when passing obliquely from one medium to another in which the velocity of sound is different.

**Acoustic resistance.** that component of the acoustic impedance which is responsible for the dissipation of energy due to friction between molecules of the air or other medium through which sound travels. Measured in acoustic ohms; analogous to electrical resistance.

**Acoustic resonance.** an increase in sound intensity as reflected waves and direct waves combine in phase. May also be due to the natural vibration of air columns or solid bodies at a particular audio frequency.

**Acoustic resonator.** an enclosure that intensifies those audio frequencies at which the enclosed air is set into natural vibration.

**Acoustic scattering.** the irregular reflection, refraction, or diffraction of a sound wave in many directions.

**Acoustic system.** arrangement of components in devices designed to reproduce audio frequencies in a specified manner.

**Acoustic transmission system.** an assembly of elements adapted to the transmission of sound.

**Acoustic treatment.** use of certain sound-absorbing materials to control the amount of reverberation in a room, hall, or other enclosed space.

**Acoustic wave.** a traveling vibration by which sound is transmitted in air or other medium. The characteristics of these waves may be described in terms of change of pressure, or particle displacement, or of density.

**Acoustic wave filter.** a device designed to separate sound waves of different frequencies. (Through electroacoustic transducers, such a filter may be associated with electric circuits.)

**Acoustics.** science of production, transmission, reception, and effects of sound. Also, in a room or other locations, those characteristics that control reflections of sound waves, and thus the sound reception in the room.

**Acoustoelectric effect.** generation of an electric current in a crystal lattice by a longitudinal sound wave.

**Action.** an organ action that denotes the assembly of key contacts and couplers.

**Aeolian.** a very soft organ stop of mild string quality.

**AES.** abbreviation for Audio Engineering Society.
**AF.** abbreviation for audio frequency, a range that extends from 20 Hz to 20 kHz.
**AFC.** abbreviation for automatic frequency control, a circuit commonly used in FM receivers to compensate for frequency drift to keep the tuner "locked" to a selected station.
**Air column.** the air space within a horn or an acoustic chamber.
**AM.** amplitude modulation; a method of superimposing intelligence on an RF carrier by amplitude variation of the carrier.
**Ambient noise.** acoustic noise in a room or other location. Usually measured with a sound-level meter.
**Amplification.** magnification or enlargement.
**Amplifier.** an electronic device that magnifies or enlarges audio voltage or power signals.
**Amplitude.** also called *peak value;* the maximum value of a wave form (with respect to one polarity).
**Anechoic enclosure.** a low-reflection audio-frequency enclosure.
**Anechoic room.** a room in which reflected sound energy is negligible; used for measurement of speaker and microphone characteristics.
**Anode.** the electrode at which electrons leave a device to enter the external circuit.
**Arpeggio.** technique of playing the notes of a chord in rapid sequence, instead of simultaneously; sometimes accomplished by automatic circuit action.
**Articulation.** the percentage of speech units understood by a listener.
**Attack.** related to *rise time.* The period of time during which a tone increases to full amplitude after a musical instrument starts to emit a tone.
**Attenuation.** opposite of amplification; reduction of audio voltage or power.
**Audio.** pertaining to frequencies corresponding to a normally audible sound wave. These frequencies range approximately from 15 Hz to 20 kHz.
**Audio level meter.** an instrument that measures audio-frequency power with reference to a predetermined level. Usually calibrated in decibels.
**Audiophile.** one who enjoys experimenting with high-fidelity equipment and who is likely to seek the best possible reproduction.
**Autotransformer.** a transformer designed with a single, tapped winding that serves as both primary and secondary.
**Background noise.** noise inherent in any electronic system.
**Backloaded horn.** a speaker enclosure arrangement in which the sound from the front of the cone feeds directly into the room, while the sound from the rear feeds into the room via a folded horn.
**Back loading.** a form of horn loading particularly applicable to low-frequency speakers; the rear radiating surface of the speaker feeds the horn and the front part of the speaker is directly exposed to the room.
**Baffle.** a partition or enclosure in a speaker cabinet that increases the length of the air path from the front to the rear radiating surfaces of the speaker.
**Baroque.** baroque music is a basic form of composition characterized by ornamentation and powerful climaxes.
**Bass.** the lower or pedal tones provided by an organ.

## Acoustic, Audio-Frequency, and Sound Terms

**Bass-reflex enclosure.** a speaker cabinet enclosure in which a portion of the radiation from the rear of the cone is channeled to reinforce the bass tones.

**Bass response.** the extent to which a speaker or audio amplifier processes low audio frequencies.

**Bassy.** a term applied to sound reproduction in which the low-frequency tones are overemphasized.

**Beat.** a successive rising and falling of a wave envelope due to alternate reinforcements and cancellations of two or more component frequencies.

**Binaural.** a type of sound recording and reproduction. Two microphones, each representing one ear and spaced about 6 in. apart, are used to pick up the sound energy to be recorded on separate tape channels. Playback is accomplished through separate amplifiers (or a two-channel amplifier) or special headphones wired for binaural listening.

**Blocked impedance.** the input impedance of a transducer when its output is connected to a load of infinite impedance.

**Blocked resistance.** resistance of an audio-frequency transducer when its moving elements are restrained so that they cannot move.

**Boffle.** a Hartley speaker enclosure that contains a group of stretched, resilient, sound-absorbing screens.

**Bourdon.** a low-pitched wood-flute organ pipe.

**Brass.** a generalized term that denotes tones resembling those from brass instruments, such as the tuba, trumpet, or cornet.

**Bridge.** a precision electrical instrument for the measurement of resistance, capacitance, and inductance values.

**Buffer.** a device, such as an electron tube or transistor, employed between an ac source and its load, principally for the purpose of isolation.

**Bus bar.** a bare electrical conductor that connects to various tone sources, or that distributes voltages to various points in an organ system.

**Cabinet, tone.** a speaker enclosure designed for operation with an electronic organ.

**Capacitor.** (obs: *condenser*). Any device designed for storage of electrostatic field energy.

**Capstan.** the spindle or shaft of a tape transport mechanism that pulls the tape past the heads.

**Capture ratio.** an FM tuner's ability to reject unwanted co-channel signals. If an undesired signal is more than 2.2 dB lower than a desired signal, the undesired signal will be completely rejected.

**Cardioid pattern.** a heart-shaped directional pickup pattern for a microphone that assists in reducing background noise.

**Carillon.** a bell-tower voice actuated from an organ keyboard; the bell tones are electronically generated.

**Cartridge.** a transducer device used with a turntable to convert mechanical channels in a disc into electrical impulses.

**Celeste.** an organ stop characterized by a slow beat of 3 or 4 Hz; it is used in the upper register, usually in the diapason family.

**Cent.** an interval between two tones, with a value of approximately $1/100$ semitone.

**Ceramic.** a piezoelectric element that is used as the basis of some phonograph pickups; it generates a potential difference when stressed or strained.

**Changer.** a record-playing device that automatically accepts and plays up to 10 or 12 discs sequentially.

**Channel.** a complete sound path. A monophonic system has one channel, a stereophonic system has two, and a quadraphonic system has four. Monophonic material may be played through a stereophonic system, and quadraphonic material may be played through a stereophonic system. An amplifier may have several inputs, such as microphone(s), tuner; mono, stereo, and quad tape; and phono.

**Channel balance.** equal response from left and right channels of a stereo amplifier. A balance control in a stereo amplifier permits adjustment for uniform sound volume from both speakers or a hi-fi system.

**Chassis.** metal frame or box that houses the circuitry of an electronic unit or system.

**Chimes.** a bell-like tone produced by striking metal tubes or rods with a hammer, or by an equivalent electronic synthesis.

**Choir.** an organ voice produced by blending several tones (of the same family) that have practically the same pitch, but differing phases. Sometimes, a choir effect is simulated by blending several tones with phase differences produced by frequency modulating one or more of the tones.

**Chord.** a combination of harmonious tones that are sounded simultaneously.

**Chord coupling.** an organ coupling mode wherein all tones for a specific chord can be played by depressing a single button or key.

**Chord organ.** an organ arranged for playing a variety of chords in harmony with solo tones. Each chord is played by depressing a single button or key.

**Chorus effect.** same as *choir*.

**Chromatic keyboard.** a keyboard with the black notes placed at the same height as the white notes, and with the same widths, to facilitate playing of chromatic scales.

**Chromatic percussion.** percussive effects that are applied to notes of an organ to simulate struck strings, plucked strings, marimba, or xylophone voices.

**Chromatic scale.** a scale composed entirely of half-steps.

**Cipher.** a tone that sounds when no key is depressed, owing to malfunction.

**Clarinet.** an organ stop for a voice that simulates clarinet tones.

**Clavier.** any keyboard or pedal board operated with either the hands or feet. A hand-operated clavier is more often termed a *manual*.

**Compensator.** a fixed or variable circuit built into a preamplifier that compensates for bass and treble alterations that were made during the recording process.

**Complex tone.** an audio wave form composed of a fundamental frequency and a number of integrally related harmonic frequencies (a pitch and a number of related overtones).

**Compliance.** physical freedom from rigidity that permits a stylus to track a record groove precisely, or of a speaker to respond to an audio signal precisely.

**Concordant.** a series of musically meaningful tones.

**Cone.** the diaphragm that sets the air in motion to generate a sound wave in a direct-radiator speaker; usually conical in shape.

**Conical horn.** a horn, the cross section of which increases as the square of its axial length.

**Console.** a cabinet that houses an electronic organ.

**Contra.** when prefixed to the name of a musical instrument, this term signifies that the tones have been lowered one octave.

**Cornopean.** an organ voice with a rich and hornlike tone color.

**Counterbass.** also termed *contrabass;* this term denotes a second bass note that will harmonize with a particular chord.

**Coupler.** a stop or tab that permits the tones on one manual of an organ to be played with the tones of another manual, or that permits the sounding of octavely related tones on the same manual.

**cps.** abbreviation for cycles per second; *see* hertz, cycle, and cycles per second.

**Crescendo.** a pedal or equivalent control for an electronic organ that rapidly brings all stops into play; an increase in voice output to maximum power capability.

**Crossover distortion.** distortion that occurs in a push–pull amplifier at the points of operation where the signals cross over the zero axis.

**Crossover frequency.** in reference to electrical dividing networks, the audio frequency at which equal power is delivered to each of the channels or speakers.

**Crossover network.** filtering circuit that selects and passes certain ranges of audio frequencies to the speakers that are designed for the particular ranges.

**Crosstalk.** in stereo high-fidelity equipment, crosstalk signifies the amount of left-channel signal that leaks into the right channel, and vice versa.

**Crystal.** a natural piezoelectric element that is used in some phono pickup cartridges and microphones.

**Crystal loudspeaker.** a speaker in which piezoelectric action is used to produce mechanical displacement. Also termed a piezoelectric loudspeaker.

**Cycle.** one complete reversal of an alternating current, including a rise to maximum in one direction, a return to zero, a rise to maximum in the other direction, and another return to zero. The number of cycles occurring in 1 s is defined as the frequency of an alternating current. The word *cycle* is commonly interpreted to mean cycles per second, in which case it is a measure of frequency. The preferred term is hertz.

**Cycles per second.** an absolute unit for measuring the frequency or "pitch" of a sound, various forms of electromagnetic radiation, and alternating electric current. *See* hertz.

**Cymbal.** a high-pitched metallic organ stop that simulates the metallic clashing sound of orchestra cymbals.

**Damping.** prevention of vibrations, response, or resonances that would cause distortion if unchecked. Mechanical control is by friction; electrical control is by resistance.

**Damping factor.** 1. For any underdamped motion during any complete oscillation, the quotient obtained by dividing the logarithmic decrement by the time required by the oscillation. 2. Numerical quantity indicating ability of an amplifier to operate a speaker properly. Values over 4 are usually considered satisfactory. 3. The ratio of rated load impedance to the internal impedance of an amplifier.

**Dead end.** the end of a sound studio with the greater sound-absorption characteristic.

**Dead room.** a room for testing the acoustic efficiency or range of electroacoustic devices such as speakers and microphones. The room is designed with an absolute minimum of sound reflection, and no two dimensions of the room are the same. The walls, floor, and ceiling are lined with sound-absorbent material.

**Decay.** a period of time over which a tone decreases from peak volume to inaudibility. It is characterized as an exponential function that defines the natural law of decay (and growth).

**Decibel (dB).** a unit for measuring relative power levels. One decibel is equal to $\frac{1}{10}$ bel, and is about the smallest change that can be detected by a critical listener.

**Deemphasis.** an attenuation of certain frequencies; in playback equalization, deemphasis offsets the preemphasis given to the higher frequencies during the recording process.

**Delay line.** an electromechanical line (or equivalent) for delaying a signal or impulse in passage between the input and output terminals; often terminated in comparatively high or low impedances, to obtain energy reflections (reverberation).

**Diapason.** the basic tone color of traditional organ voices, as produced by open or stopped pipes.

**Diaphragm.** thin, flexible sheet that vibrates when struck by sound waves, as in a microphone, or that produces sound waves when moved back and forth at an audio-frequency rate, as in a headphone or a speaker.

**Diffracted wave.** a sound wave that has struck an object and has been bent or deflected, other than by reflection or refraction.

**Diffraction.** the bending of sound waves as they pass through an object or barrier, thereby producing a diffracted wave. Also, the phenomenon whereby waves traveling in straight paths bend around an obstacle.

**Diode.** a unilateral electronic device that is used in rectification, wave shaping, switching, and other circuit applications.

**Direct-coupled amplifier.** a dc amplifier in which the output of one stage is coupled to the input of the next stage by a direct connection or a low-value resistor.

**Direct coupling.** the association of two or more circuits by means of an inductance, a resistance, a wire, or a combination of these so that both direct and alternating currents can be coupled.

**Directivity factor.** of a transducer used for sound emission, the ratio of the intensity of the radiated sound at a remote point in a free field on the principal axis to the average intensity of the sound transmitted through a sphere passing through the remote point and concentric with the transducer.

**Directivity index.** also termed directional gain. A measure of the directional properties of a transducer. It is the ratio in decibels of the average intensity over the whole sphere surrounding the projector to the intensity on the acoustic axis.

**Direct-radiator speaker.** a speaker in which the radiating element acts directly on the air instead of relying on any other element, such as a horn.

**Discordant.** tones that are unrelated by established principles of harmony.

**Distortion.** deviations from an original sound that occur in the reproduction process. Harmonic distortion disturbs the original relationship between a tone and other tones naturally related to it. Intermodulation distortion introduces new tones that result from the beating of two or more original tones.

**Divider.** a circuit, device, or arrangement that reduces a signal voltage to a certain fraction of its input value, or that generates a subharmonic of an input signal frequency.

**Dividing network.** same as *crossover network*.

**Doppler tone cabinet.** a tone-cabinet design in which one or more speakers are rotated or in which a baffle is rotated to produce a mechanical vibrato/tremolo effect.

**Double touch.** a key-contact design for an electronic organ that provides actuation of an additional circuit when somewhat more than normal finger pressure is applied.

**Doubling.** the generation of a large amount of second-harmonic distortion owing to a nonlinear motion of a speaker cone.

**Drone cone.** an undriven speaker cone mounted in a bass-reflex enclosure.

**Ducted port.** a form of bass-reflex speaker enclosure in which a tube is mounted behind the reflex port.

**Dulciana.** a flute voice with a small and slightly stringy tone.

**Dynamic cartridge (electrodynamic).** a magnetic phono pickup in which a moving coil in a magnetic field generates voltages to form an audio signal.

**Dynamic microphone.** a microphone that operates on the same basic principle as a dynamic cartridge.

**Dynamic speaker.** also termed a moving-coil speaker. The moving diaphragm is attached to a coil, which is conductively connected to the source of electric energy and placed in a constant magnetic field. The current through the coil interacts with the magnetic field, causing the coil and diaphragm to move back and forth in accordance with the current variations through the coil.

**Dyne per square centimeter.** the unit of sound pressure. Originally called a bar, but now termed by the full expression.

**Eccles–Jordan oscillator.** also termed a flip-flop, or bistable multivibrator; used for frequency division in electronic organ networks.

**Echo.** a delayed repetition (sometimes several rapid repetitions) of the original sound.

**Effective current.** the value of alternating or varying current that will produce the same amount of heat as the same value of direct current. Also called rms current.

**Effective sound pressure.** the root mean square of the instantaneous sound pressure at one point over a complete cycle. The unit is the dyne per square centimeter.

**Efficiency.** in a speaker, the ratio of power applied to the input terminals expressed as a percentage.

**Electroacoustic.** pertaining to a device, as a speaker, that involves both electric current and sound-frequency pressures.

**Electroacoustic transducer.** a device that receives excitation from an electric system and delivers its output to an acoustic system, or vice versa.

**Electrodynamic speaker.** a speaker consisting of an electromagnet, termed the *field coil,* through which a direct current flows.

**Electromagnetic.** pertaining to a phenomenon that involves the interaction of electric and magnetic field energy.

**Electrostatic speaker.** a type of speaker in which sound is produced by charged plates that are caused to move while one is changed from positive to negative polarity, resulting in forces of attraction or repulsion.

**Electrostatic tweeter.** a speaker with a movable flat metal diaphragm and a nonmovable metal electrode capable of reproducing high audio frequencies. The diaphragm is driven by the varying high voltage that is applied to the plates.

**Enclosure.** a housing that is acoustically designed for a speaker or speakers. Also called a tone cabinet in electronic organ technology.

**Equal-loudness contours.** Fletcher–Munson curves, *q.v.*

**Erase head.** the leadoff head in a tape recorder that erases previous recordings from the passing tape by generating a strong and random magnetic field.

**Excess sound pressure.** the total instantaneous pressure at a point in a medium containing sound waves, minus the static pressure when no sound waves are present. The unit is the dyne per square centimeter.

**Expression control.** an organ volume control, usually operated with the right foot.

**Extended octave.** a tone above or below a note on a standard keyboard that sounds when a specific coupler is actuated.

**Fast decay.** a rapid attenuation of a tone after its keyswitch has been released.

**Feed reel.** the reel in a tape recorder that supplies the tape.

**FET (field-effect transistor).** a transistor of the voltage-operated-device classification, instead of the current-operated type as a bipolar transistor.

**Fidelity.** the faithfulness of sound reproduction.

**Filter network.** a reactive network that is designed to provide specified attenuation to signals within certain frequency limits; basic filters are termed low-pass, high-pass, bandpass, and band-reject deigns.

**Flare factor.** a number that expresses the degree of outward curvature of a speaker horn.

**Flat.** a note that is a half-step or semitone lower than its related natural pitch.

**Flat response.** a characteristic of an audio system whereby any tone is reproduced without deviation in intensity for any part of the frequency range that it covers.

**Fletcher–Munson curves.** also called equal-loudness contours. A group of sensitivity curves showing the characteristics of the human ear for different intensity levels between the threshold of hearing and the threshold of feeling. The reference frequency is 1 kHz.

**Flute.** a basic electronic organ tone color that simulates the orchestral flute.

**Flutter.** a form of distortion caused when a tape transport or a turntable is subject to rapid speed variation.

**FM.** frequency modulation.

**FM sensitivity.** the minimum input signal required in an FM receiver to produce a specified output signal having a specified signal-to-noise ratio.

**FM stereo.** broadcasting over FM frequencies of two sound signals within a single channel. A *multiplexing* technique is utilized.

**Folded horn.** a type of speaker enclosure that employs a horn-shaped passageway that improves bass respone.

**Force factor (of an electroacoustic transducer).** the complex quotient of the force required to block the mechanical or acoutic system, divided by the corresponding current in the electrical system. The complex quotient of the resultant open-circuit voltage in the electric system divided by the velocity in the mechanical or acoustic system.

**Force-summing device.** in a transducer, the element directly displaced by the applied stimulus.

**Formant filter.** a wave-shaping network or device that changes the waveform of a tone-generator signal into a desired musical tone waveform.

**Forte.** a forte tab (solo tab) increases the volume of other tabs that are depressed at the time; a forte tab has no voice of its own.

**Foundation voice.** a definitive organ voice, such as the diapason and dulciana voices.

**Free impedance.** also called *normal* impedance. The input impedance of a transducer when the load impedance is zero.

**Free motional impedance.** the complex remainder after the blocked impedance of a transducer has been subtracted from the free impedance.

**Free-running oscillator.** an oscillator that generates an output in the absence of a synchronizing signal or a trigger signal.

**Free sound field.** a field in a medium free from discontinuities or boundaries. In practice, it is a field in which the boundaries cause negligible effects over the region of interest.

**Frequency.** the number of complete vibrations or cycles completed in 1 s by a waveform, and measured in hertz.

**Frequency modulation.** a method of broadcasting that varies the frequency of the carrier instead of its amplitude. FM is the selected high-fidelity medium for broadcasting high-quality program material.

**Frequency range.** the limiting values of a frequency spectrum, such as 20 Hz to 20 kHz.

**Frequency response.** the frequency range over which an audio device or system will produce or reproduce a signal within a certain tolerance, such as ±1 dB.

**Fundamental.** the normal pitch of a musical tone; usually, the lowest frequency component of a tonal waveform.

**Gain.** the value of amplification that a signal obtains in passage through an amplifying stage or system.

**Gate circuit.** a circuit that operates as a selective switch and permits conduction over a specified interval.

**Gemshorn.** a flute organ voice with a bright tone color.

**Generator.** a tone or signal source, such as an oscillator, frequency divider, or magnetic tone wheel.

**Glide.** also termed *glissando*. A rapid series of tones produced by a slight shift in pitch of successive tones.

**Glockenspiel.** also called *orchestra bells*. An electromechanical arrangement that stimulates the bells used in orchestras.

**Great manual.** also called *accompaniment manual* or *lower manual*. A keyboard used for playing the accompaniment to a melody.

**Grill.** a decorative and protective sound-transparent structure and/or mesh that forms the front surface of a speaker enclosure.

**Half-tone.** also called *semitone*. The relation between adjacent pitches on the tempered scale.

**Harmonic.** a frequency component of a complex waveform that bears an integral relation to the fundamental frequency. Also called *overtone*.

**Harmonic distortion.** *see* distortion.

**Harmony.** musical support for a melody, consisting of two or more notes played simultaneously.

**Head.** electromagnetic device used in magnetic tape recording to convert an audio signal to a magnetic pattern, and vice versa.

**Headphones.** small sound reproducers resembling miniature speakers used either singly or in pairs, usually attached to a headband to hold the phones snugly against the ears. Available in monophonic or stereophonic design.

**Helmholtz resonator.** an acoustic enclosure with a small opening that causes the enclosure to resonate. The frequency of resonance is a function of the resonator geometry.

**Hertz.** a unit of frequency equal to 1 cycle per second (cps).

**High fidelity.** the characteristic that enables an audio system to reproduce sound as nearly like the original as possible.

**Hole-in-the-middle effect.** the lower volume or absence of sound between the left and right speakers of a stereo system.

**Horn.** also called an acoustic horn. A tubular or rectangular enclosure for radiation of acoustic waves.

**Horn cutoff frequency.** a frequency below which an exponential horn will not function correctly because it fails to provide for proper expansion of the sound waves.

**Horn loading.** a method of coupling a speaker diaphragm of the listening space by an expanding air column that has a small throat and a large mouth.

**Horn mouth.** the wide end of a horn.

**Horn speaker.** a speaker in which a horn couples the radiating element to the medium.

**Horn throat.** the narrow end of a horn.

**Hum.** noise generated in an audio or other electronic device by a source or sources of electrical disturbance.

**IC.** abbreviation for *Integrated Circuit*. Integral solid-state units that include transistors, resistors, semiconductor diodes, and often capacitors, all of which are formed simultaneously during fabrication.

# Acoustic, Audio-Frequency, and Sound Terms

**Ideal transducer.** theoretically, any linear passive transducer that, if it dissipated no energy and, when connected to a source and load, presented its combined impedance to each, would transfer maximum power from source to load.

**IHFM (IHF).** refers to the Institute of High Fidelity Manufacturers, now called the Institute of High Fidelity, Inc. This group devises and publishes standards and ratings for high-fidelity equipment.

**Image rejection.** the ability of a receiver to reject interference that is produced by an undesired input frequency that beats with the local-oscillator frequency to produce an abnormal IF frequency.

**Impedance.** an electrical unit, expressed in ohms, that denotes the amount of opposition to ac flow by a device or a circuit.

**Infinite baffle.** a speaker mounting arrangement in which the front and back waves from a cone are totally isolated from each other.

**Input.** connection through which an electric current is fed into a device, circuit, or system.

**Intermodulation distortion (IM).** two distinct and separate test frequencies are mixed in an amplifier, and their difference-frequency output is measured in IM percentage. *See* distortion

**Interval.** the difference in pitch between two musical tones.

**Jack.** a female receptacle for a plug-type connector.

**Keybed.** a shelf or horizontal surface on which a keyboard is mounted.

**Keyboard.** a bank of keys, comprising black and white sets, arranged in ascending tones.

**Keynote.** the tonic or first note of a particular scale.

**Keyswitch.** a switch that closes when a key is depressed, thereby actuating a tone generator.

**Kinura.** a reed stop that has dominant harmonics and a subordinate fundamental.

**Labyrinth.** a speaker enclosure with absorbing air chambers at the rear to eliminate acoustic standing waves.

**Lateral systems.** a system of disc recording in which a stylus moves from side to side (laterally).

**Leslie speaker.** a generic term, originally a trade name, denoting a tone cabinet with a mechanical vibrato/tremolo assembly.

**Level indicator.** a neon bulb, meter, or "eye" tube used to indicate recording levels.

**Lissajous figures.** an $XY$ plot of voltage or current phase relations, usually produced automatically on the screen of a cathode-ray tube.

**Load.** a device, circuit, or system that absorbs or converts power from an electrical source, as a speaker converts power from an amplifier.

**Loudness control.** an audio-frequency filtering arrangement that boosts the treble and particularly the bass tones in an amplifier as the volume level is reduced; it compensates for the listener's reduced sensitivity to tones at the extreme ends of the audio range at low volume levels.

**Loudspeaker.** equivalent term for speaker.

**Loudspeaker dividing network.** equivalent term for crossover network.
**Loudspeaker impedance.** equivalent term for speaker impedance.
**Loudspeaker system.** equivalent term for speaker system.
**Louver.** the grille of a speaker.
**Magnetic armature speaker.** a speaker comprising a ferromagnetic armature actuated by magnetic attraction.
**Magnetic speaker.** a speaker in which acoustic waves are produced by mechanical forces resulting from magnetic reaction.
**Magnetic tape.** plastic tape with an iron-oxide coating for magnetic recording.
**Manual.** *see* clavier.
**Manual player.** manual record-playing device used with a changer-type machine.
**Master oscillator.** a source of a tone signal; it may be utilized directly, or it may be processed through one or more frequency dividers; these are also oscillators, but are of the driven type.
**Mean free path.** the average distance that sound waves travel between successive reflections in an enclosure.
**Mechanicals.** organ effects that are not voices in the strict sense of the term; thus, forte (solo), percussion effects, and couplers.
**Mechanical tone generator.** a mechanical unit for generation of electrical impulses that are subsequently converted into audible tones.
**Megohm.** a multiple unit that denotes 1 million $\Omega$.
**Mel.** a unit of pitch; a simple 1-kHz tone, 40 dB above a listener's threshold, produces a pitch of 1,100 mels. The pitch of any sound that is judged by the listener to be $n$ times that of a 1-mel tone is denoted as $n$ mels.
**Melodia.** an organ solo stop of the flute family.
**Melody.** also called a tune; usually played sequentially note by note on the swell or solo manual.
**Micro.** a prefix that denotes one millionth.
**Microbar.** a unit of pressure commonly used in acoustics. One microbar is equal to 1 dyne per $cm^2$.
**Milli.** a prefix that denotes one thousandth.
**Mixing.** a blend of two or more electrical signals or acoustic waves.
**Modulation.** a process wherein low-frequency information is encoded into a higher-frequency carrier or subcarrier; subdivisions include amplitude, frequency, and phase modulation, with various combinations and derivatives thereof.
**Monophonic.** a recording and reproduction system in which all program material is processed in one channel.
**Monorange speaker.** a speaker that provides the full spectrum of audio frequencies.
**Moving-coil speaker.** also termed a *dynamic speaker*. A speaker in which the moving diaphragm is attached to a coil, which is driven by audio-frequency currents. These currents interact with a fixed magnetic field and cause the diaphragm to vibrate in unison.
**Multiplexing.** a system of broadcasting in which two or more separate channels are transmitted on one FM carrier, as in stereophonic broadcasting.

**Multivibrator.** a relaxation oscillator, usually developing a semisquare wave form. Subclassifications include the astable, monostable, and bistable types.

**Muting.** a silencing process or action.

**NAB curve.** tape-recording equalization curve established by the National Association of Broadcasters.

**Nazard.** an organ voice that stimulates a piccolo-type pipe-organ voice.

**Near field.** the acoustic radiation field close to the speaker or some other acoustic source.

**Neon lamp.** a gas diode that emits an orange glow, and operates as an indicator, protective switch, regulator, relaxation oscillator, or divider.

**Network.** a comparatively elaborate electrical or electronic circuit arrangement.

**Nonchromatic percussion.** a percussion effect that has no dominant pitch, such as wood-block, drum, castanet, or cymbal effects.

**Note.** a single musical tone, identified by the letters A through G, plus sharp or flat superscripts.

**Octave.** a pair of tones are separated by an octave if one has twice the frequency of the other.

**Octave coupling.** an organ coupling arrangement wherein the depression of a key causes another note an octave higher or lower in pitch to sound simultaneously.

**Ohm.** the unit of electrical resistance, defined as a unitary voltage/current ratio.

**Oscillator.** an electronic, electrical, or mechanical generator of an electrical signal.

**Outphasing.** an organ voicing method wherein specified harmonics or subharmonics are added to or subtracted from a tone signal prior to its application to a formant filter. In a *chiff* outphasing circuit, certain harmonics are added to the tone signal during its attack period.

**Output.** a connection or conductor through which an electrical signal emerges from an electrical or electronic device, circuit, or system.

**Overall loudness level.** a measure of the response of human hearing to the strength of a sound. It is scaled in phons and is an overall single evaluation calculated for the level of sound pressure of several individual bands.

**Overtone.** same as *harmonic.*

**Partial.** any one of the various frequencies contained in a complex wave form that corresponds to a musical tone.

**Patch cord.** a shielded cable utilized to connect one audio device to another.

**Peak sound pressure.** the maximum absolute value of instantaneous sound pressure for any specified time interval. The most common unit is the microbar.

**Pedal.** the pedal keyboard of an electronic organ; also termed *clavier.*

**Pedal clavier.** a pedal keyboard.

**Pedal divider.** a frequency-divider section associated with the tone generators actuated by the foot pedals.

**Pedal generator.** a tone generator utilized to produce the bass notes of an organ.

**Pedal keyboard.** same as *pedal clavier*.

**Percussion.** characteristic tones, as produced by plucking or striking strings.

**Permanent-magnet speaker.** a moving-conductor speaker in which the steady magnetic field is produced by a permanent magnet.

**pF.** abbreviation for *picofarad*.

**Phase.** position occupied at any instant in its cycle by a periodic wave; a part of a sound wave or signal with respect to its passage in time. One signal is said to be in phase with, or to lead, or to lag, another reference signal.

**Phase inverter.** an amplifier that provides an output which is 180 degrees out of phase with its input, or an amplifier that provides a pair of output voltages which are 180 degrees out of phase with each other.

**Phon.** the unit for measurement of the apparent loudness level of a sound. Numerically equal to the sound-pressure level, in decibels relative to 0.0002 $\mu$bar, of a 1-kHz tone that is considered by listeners to be equivalent in loudness to the sound under consideration.

**Pickup cartridge.** a device used with a turntable to convert mechanical variations into electrical impulses.

**Picofarad.** a unit equal to 1 micro-microfarad.

**Piezoelectric speaker.** a speaker that employs a piezoelectric substance as a driver or motor.

**Piston.** an organ stop that is operated by pulling or pushing a knob. A piston generally operates groups of conventional stops.

**Piston action.** the movement of a speaker cone or diaphragm when driven at the bass audio frequencies.

**Pitch.** that characteristic of a sound which places it on a musical scale.

**Pizzicato.** an organ sound effect that simulates the rapid plucking of strings.

**Playback head.** the last head of a tape recorder, or the only head on a tape player, which converts the magnetic pattern impressed on a passing tape into an audio signal.

**Plug-type connector.** a mating connector for a jack.

**PM.** permanent magnet.

**Polyester backing.** a plastic material used as a base for magnetic recording tape.

**Port.** an opening in the baffle of a bass-reflex speaker enclosure for selective radiation of sound waves.

**Power.** a unit of the rate at which work is done, or energy is consumed, or energy is generated; electrical power is measured basically in rms watts.

**Power amplifier.** an amplifier that drives a speaker in an audio system.

**Power bandwidth.** specification of a higher-frequency limit and a lower-frequency limit for an amplifier, between which the harmonic distortion is no greater at −3 dB of maximum rated power than the harmonic distortion measured at 1 kHz and at maximum rated power output.

**Power output.** the signal power delivered by an audio amplifier, measured in watt units.

**Power supply.** a source of electrical energy; usually, an arrangement that converts alternating current into virtually pure direct current.

**Preamplifier.** amplifying arrangement that steps up a very weak input signal to a suitable level for driving an intermediate amplifier or a power amplifier.

**Preemphasis.** a deliberate exaggeration of the high-frequency components in an audio signal.

**Presence.** the quality of naturalness in sound reproduction. When the presence of a system is good, the illusion is that the sounds are being produced intimately at the speaker.

**Preset.** a control that turns on a group of voices or that turns them off without actuating any tabs.

**Print-through.** magnetization of a layer of tape by an adjacent layer.

**Pulse.** an electrical transient, or a series of repetitive surges.

**Quadraphonic.** a system whereby sound that is picked up by four separate microphones is recorded on separate channels and played back through separate channels that drive individual speakers.

**Quality.** relates to the harmonic content of a complex tonal wave form; also termed *timbre*.

**Quarter-track recorder.** a tape recorder that utilizes one quarter of the width of the tape for each recording; in stereo operation, two of the four tracks are used simultaneously.

**Quieting.** standard of separation between background noise and the program material from a tuner.

**Record head.** the second head of a tape recorder; used to convert an audio signal to a magnetic pattern on the passing tape.

**Recording amplifier.** an amplifying section in a tape recorder that prepares an audio signal for application to the record head, and bias current to the erase head.

**Record-playback head.** a head on a tape recorder that performs both recording and playback functions.

**Reed.** one of the basic tone-color groups of organ voices that simulate orchestral reeds.

**Reference acoustic pressure.** that magnitude of a complex sound that produces a sound-level meter reading equal to the reading that results from a sound pressure of 0.0002 dyne per $cm^2$ at 1 kHz. Also called *reference sound level*.

**Register.** a range of notes included by a clavier or manual; range of notes employed in playing a particular musical composition.

**Relay.** an electromagnetically operated switching device.

**Reproducer.** a device used to translate electrical signals into sound waves.

**Resultant.** denotes a tone that is produced when two notes one fifth apart and an octave higher than the desired note are sounded to produce the desired pitch; a mode of generating *synthetic bass*.

**Reverberation.** the persistence of sound due to repeated reflections from walls, ceiling, floor, furniture, and occupants in a room.

**Reverberation period.** the time required for the sound in an enclosure to decay to one millionth (60 dB) of its original intensity.

**Reverberation strength.** the difference between the level of a plane wave that produces in a nondirectional transducer a response equal to that produced

**Reverberation time.** for a given frequency, the time required for the average sound-energy density, originally in a steady state, to decay to one millionth (60 dB) of its initial value after the source is stopped.

by the reverberation corresponding to a 1-yard range from the effective center of the transducer.

**Rhythm section.** an organ section that generates nonchromatic percussion effects in a periodic manner, either automatically or manually.

**RIAA curve.** standard disc-recording curve specified by the Record Industry Association of America.

**Ribbon tweeter.** a high-frequency speaker, usually horn loaded, in which a stretched, straight flat ribbon is used instead of a conventional voice coil.

**Rolloff.** the rate at which a frequency response curve decreases in amplitude; it is usually stated in decibels per octave or decibels per decade.

**Rumble.** a low-frequency vibration originating from a vibrating electric motor in a turntable.

**Rumble filter.** a low-frequency filter circuit designed to minimize or to eliminate rumble interference.

**Sabin (square-foot unit of absorption).** a measure of the sound absorption of a surface. It is equivalent to 1 $ft^2$ of a perfectly absorptive surface.

**Scale.** a series of eight consecutive whole notes.

**Scratch filter.** a high-frequency filter circuit that minimizes scratchy sounds in playback of deteriorated discs.

**Sectoral horn.** a horn with two parallel and two diverging sides.

**Selectivity.** a measure of the ability of an electronic device to select a desired signal and to reject adjacent interfering signals; also termed *bandwidth*.

**Semitone.** the relation between adjacent ptiches on the tempered scale.

**Sensitivity.** the minimum value of input signal that is required by an electronic unit, such as a tuner, to deliver a specified output signal level.

**Separation.** the degree to which one channel's information is excluded from another channel; customarily expressed in decibel units.

**Sforzando.** a form of *crescendo*, but also employing discordant tones.

**Sharp.** removed by a semitone from a reference pitch.

**Signal-to-noise ratio.** the extent to which program material exceeds the background noise level; customarily expressed in decibel units.

**Sine wave.** graphical representation of simple harmonic motion.

**Soft-suspension speaker.** a speaker design without inherent springiness; it utilizes the reaction of a trapped back wave for restorative force.

**Solo manual.** the upper manual of a two-manual organ; also called a *swell manual*.

**Sone.** a unit of loudness. A simple 1-kHz tone, 40 dB above a listener's threshold, produces a loudness of 1 sone. The loudness of any sound that is judged by the listener to be $n$ times that of the 1-sone one is $n$ sones.

**Sound.** also called a *sound wave*. An alteration in pressure, stress, particle displacement, or velocity, propagated in an elastic medium. Also called a *sound sensation*. The auditory sensation evoked by a sound wave.

**Sound absorption.** the conversion of sound energy into some other form (usually heat) in passing through a medium or in striking a surface.

**Sound absorption coefficient.** the incident sound energy absorbed by a surface or a medium, expressed as a fraction.

**Sound pressure level.** in decibels, 20 times the logarithm of the ratio of the pressure of a sound to the reference pressure, which must be explicitly stated (usually, either $2 \times 10^{-4}$ or 1 dyne per cm$^2$). Also, the pressure of an acoustic wave stated in terms of dynes per square centimeter or microbars.

**Sound reflection coefficient.** also called *acoustical reflectivity*. Ratio at which the sound energy reflected from a surface flows on the side of incidence to the incident rate of flow.

**Sound-reproducing system.** a combination of transducers and associated equipment for reproducing prerecorded sound.

**Sound spectrum.** the frequency components included within the range of audible sound.

**Speaker.** an electroacoustic transducer that radiates acoustic power into the air.

**Speaker efficiency.** ratio of the total useful sound radiated from a speaker at any frequency to the electrical power applied to the voice coil.

**Speaker impedance.** the rated impedance of the voice coil in a speaker.

**Speaker system.** a combination of one or more speakers and all associated baffles, horns, and dividing networks used to couple together the driving electric circuit and the acoustic medium.

**Speaker voice coil.** in a moving-coil speaker, the component that is moved back and forth in response to the applied audio voltage.

**Specific acoustic impedance.** also called unit-area acoustic impedance. The complex ratio of sound pressure to particle velocity at a point in a medium.

**Specific acoustic reactance.** the imaginary component of the specific acoustic impedance.

**Specific acoustic resistance.** the real component of the specific acoustic impedance.

**Squawker.** a midrange speaker.

**Standing waves.** reflected waves that alternately cancel and reinforce at various distances.

**Stereophonic sound.** a system wherein sound energy that is picked up by two separated microphones is recorded on separate channels and is then played back through separate channels that drive individual speakers.

**Stop.** a tab or other switch form that selects and/or mixes various voices and footages in an electronic organ system.

**Strength of a simple sound source.** the rms magnitude of the total air flow at the surface of a simple source in cubing meters per second, where a simple source is taken to be a spherical source, the radius of which is small compared with one-sixth wavelength.

**Strength of a sound source.** the maximum instantaneous rate of volume displacement produced by the source when emitting a sinusoidal wave.

**String.** one of the four basic tone-color groups that simulates orchestral strings.

**Stroboscopic disc.** a cardboard or plastic disc with a specialized printed design suitable for checking turntable speed.

**Stylus.** same as phonograph needle.

**Subharmonic.** an integral submultiple of the fundamental frequency in a tonal waveform.

**Supertweeter.** a speaker designed to reproduce the highest frequencies in the audio range.

**Sustain.** an effect produced by a note that diminishes in intensity gradually after the key has been released.

**Swell manual.** the upper manual of an organ; also termed the *solo manual*.

**Synthetic bass.** a method of bass-tone generation that depends on the nonlinear response of the ear; pertinent harmonics are intensified, and although the fundamental is not present, the listener obtains the impression that the tone is complete.

**T- pad.** a three-element fixed attenuator.

**Tablet (tab).** a rocker-type switch control that selects an organ voice or footage.

**Take-up reel.** a reel on a tape recorder that winds the tape after it passes the heads.

**Tape deck.** a tape unit without a power supply or speaker.

**Temperament.** a mode of tuning an instrument scale so that successive tones correspond to specified intervals.

**Tempered scale.** an arrangement of musical pitches such that successive notes have equal frequency ratios.

**Terminal.** electrical connection point.

**Tribia.** an organ voice that simulates flute tones.

**Timbre.** also termed *tone color*. The distinguishing quality of a sound that depends primarily upon harmonic content and secondarily upon volume.

**Tone.** the fundamental frequency or pitch of a musical note.

**Tone arm.** a pivoted arm on a turntable that houses the pickup cartridge.

**Tone burst.** a test signal comprising short sequences of sine-wave energy.

**Tone color.** also termed *timbre*. Classified as *diapason, flute, string,* or *reed*.

**Tone control.** a control that provides variation of an amplifier's frequency response.

**Tone generator.** an organ section that generates the basic voice wave forms.

**Tracking.** the path of a phono stylus within the grooves of a disc.

**Transducer.** a device that converts one form of energy into another form.

**Transient.** an electrical surge.

**Transient response.** the ability of a speaker to follow sudden changes in signal level.

**Tremolo.** an amplitude modulation of a tone at a rate of approximately 7 Hz.

**Triaxial speaker.** a dynamic speaker unit consisting of three independently driven units combined into a single speaker.

**Tuning fork.** a precision source of pitch, usually designed as a U structure, and supported at its nodal point.

**Turntable.** same as *record player*.

**Tweeter.** a speaker designed to reproduce the higher audio frequencies, usually those above 3,000 Hz.

**Acoustic, Audio-Frequency, and Sound Terms** 227

**Varistor.**  a voltage-dependent resistor.
**Vibrato.**  a frequency modulation of a tone at a rate of approximately 7 Hz.
**Voice.**  an organ tone of specified timbre.
**Volume.**  same as *expression*. A relative sound level.
**Watt.**  a power unit, equal to the product of 1 V and 1 A.
**Woofer.**  a speaker designed to reproduce bass tones.
**Wow.**  a form of distortion that occurs when a magnetic tape varies back and forth in speed, or a turntable varies similarly in rpm.
**Wow-wow.**  a very slow vibrato effect.

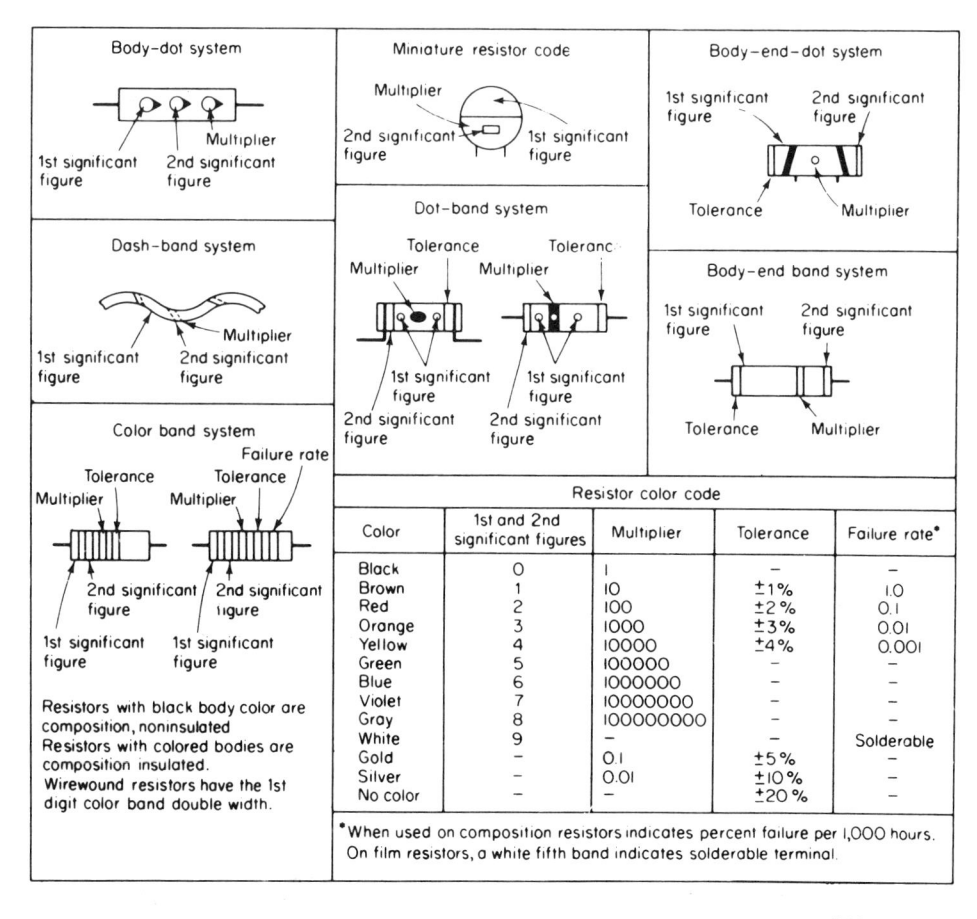

# appendix one
# RESISTOR COLOR CODE

### Body-dot system
1st significant figure
2nd significant figure
Multiplier

### Miniature resistor code
Multiplier
2nd significant figure
1st significant figure

### Body-end-dot system
1st significant figure
2nd significant figure
Tolerance
Multiplier

### Dash-band system
1st significant figure
2nd significant figure
Multiplier

### Dot-band system
Tolerance
Multiplier
1st significant figure
2nd significant figure
Tolerance
Multiplier
1st significant figure
2nd significant figure

### Body-end band system
1st significant figure
2nd significant figure
Tolerance
Multiplier

### Color band system
Tolerance
Multiplier
2nd significant figure
1st significant figure
Failure rate
Tolerance
Multiplier
2nd significant figure
1st significant figure

Resistors with black body color are composition, noninsulated.
Resistors with colored bodies are composition insulated.
Wirewound resistors have the 1st digit color band double width.

### Resistor color code

| Color | 1st and 2nd significant figures | Multiplier | Tolerance | Failure rate* |
|---|---|---|---|---|
| Black | 0 | 1 | – | – |
| Brown | 1 | 10 | ±1% | 1.0 |
| Red | 2 | 100 | ±2% | 0.1 |
| Orange | 3 | 1000 | ±3% | 0.01 |
| Yellow | 4 | 10000 | ±4% | 0.001 |
| Green | 5 | 100000 | – | – |
| Blue | 6 | 1000000 | – | – |
| Violet | 7 | 10000000 | – | – |
| Gray | 8 | 100000000 | – | – |
| White | 9 | – | – | Solderable |
| Gold | – | 0.1 | ±5% | – |
| Silver | – | 0.01 | ±10% | – |
| No color | – | – | ±20% | – |

*When used on composition resistors indicates percent failure per 1,000 hours. On film resistors, a white fifth band indicates solderable terminal.

## appendix two
# CAPACITOR COLOR CODE

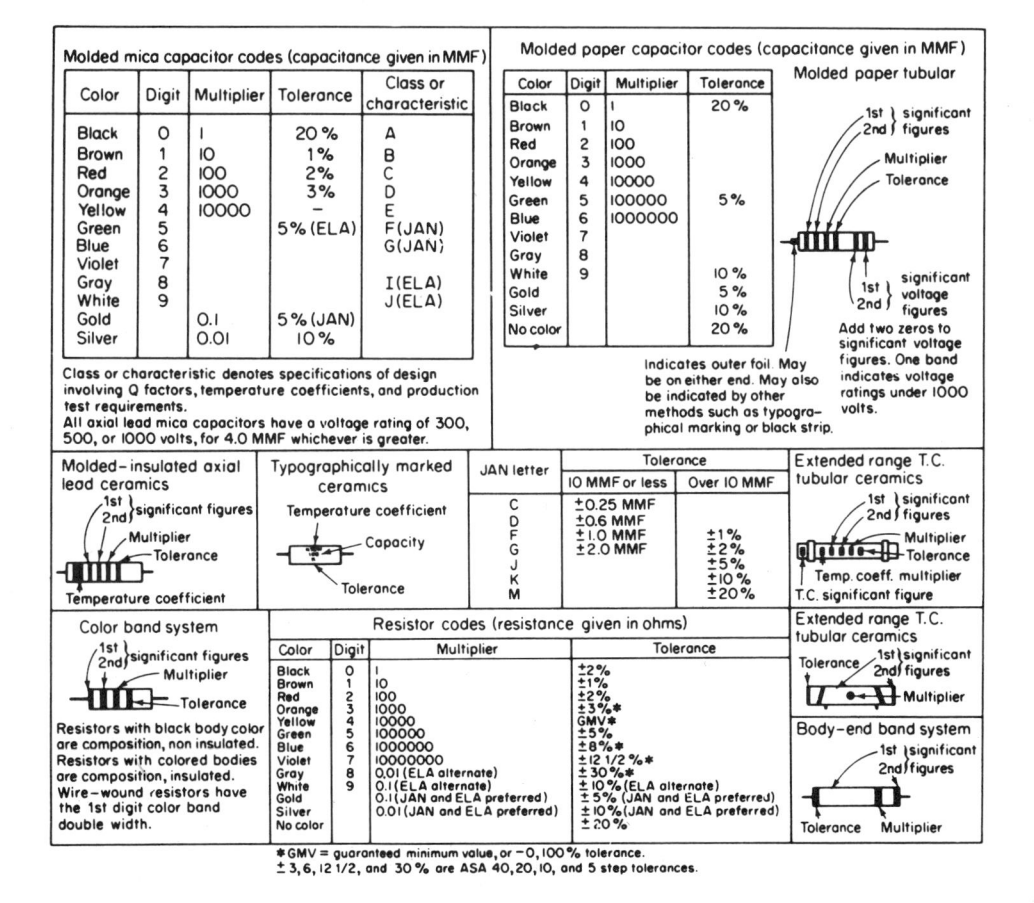

### Molded mica capacitor codes (capacitance given in MMF)

| Color | Digit | Multiplier | Tolerance | Class or characteristic |
|---|---|---|---|---|
| Black | 0 | 1 | 20% | A |
| Brown | 1 | 10 | 1% | B |
| Red | 2 | 100 | 2% | C |
| Orange | 3 | 1000 | 3% | D |
| Yellow | 4 | 10000 | – | E |
| Green | 5 | | 5% (ELA) | F (JAN) |
| Blue | 6 | | | G (JAN) |
| Violet | 7 | | | |
| Gray | 8 | | | I (ELA) |
| White | 9 | | | J (ELA) |
| Gold | | 0.1 | 5% (JAN) | |
| Silver | | 0.01 | 10% | |

Class or characteristic denotes specifications of design involving Q factors, temperature coefficients, and production test requirements.
All axial lead mica capacitors have a voltage rating of 300, 500, or 1000 volts, for 4.0 MMF whichever is greater.

### Molded paper capacitor codes (capacitance given in MMF)

| Color | Digit | Multiplier | Tolerance |
|---|---|---|---|
| Black | 0 | 1 | 20% |
| Brown | 1 | 10 | |
| Red | 2 | 100 | |
| Orange | 3 | 1000 | |
| Yellow | 4 | 10000 | |
| Green | 5 | 100000 | 5% |
| Blue | 6 | 1000000 | |
| Violet | 7 | | |
| Gray | 8 | | |
| White | 9 | | 10% |
| Gold | | | 5% |
| Silver | | | 10% |
| No color | | | 20% |

**Molded paper tubular**
1st, 2nd: significant figures
Multiplier
Tolerance
Indicates outer foil. May be on either end. May also be indicated by other methods such as typographical marking or black strip.

Add two zeros to significant voltage figures. One band indicates voltage ratings under 1000 volts.

### Molded-insulated axial lead ceramics
1st, 2nd: significant figures
Multiplier
Tolerance
Temperature coefficient

### Typographically marked ceramics
Temperature coefficient
Capacity
Tolerance

### JAN letter tolerance

| JAN letter | 10 MMF or less | Over 10 MMF |
|---|---|---|
| C | ±0.25 MMF | |
| D | ±0.6 MMF | |
| F | ±1.0 MMF | ±1% |
| G | ±2.0 MMF | ±2% |
| J | | ±5% |
| K | | ±10% |
| M | | ±20% |

### Extended range T.C. tubular ceramics
1st, 2nd: significant figures
Multiplier
Tolerance
Temp. coeff. multiplier
T.C. significant figure

### Color band system
1st, 2nd: significant figures
Multiplier
Tolerance

Resistors with black body color are composition, non insulated.
Resistors with colored bodies are composition, insulated.
Wire-wound resistors have the 1st digit color band double width.

### Resistor codes (resistance given in ohms)

| Color | Digit | Multiplier | Tolerance |
|---|---|---|---|
| Black | 0 | 1 | ±2% |
| Brown | 1 | 10 | ±1% |
| Red | 2 | 100 | ±2% |
| Orange | 3 | 1000 | ±3%* |
| Yellow | 4 | 10000 | GMV* |
| Green | 5 | 100000 | ±5% |
| Blue | 6 | 1000000 | ±8%* |
| Violet | 7 | 10000000 | ±12 1/2%* |
| Gray | 8 | 0.01 (ELA alternate) | ±30%* |
| White | 9 | 0.1 (ELA alternate) | ±10% (ELA alternate) |
| Gold | | 0.1 (JAN and ELA preferred) | ±5% (JAN and ELA preferred) |
| Silver | | 0.01 (JAN and ELA preferred) | ±10% (JAN and ELA preferred) |
| No color | | | ±20% |

*GMV = guaranteed minimum value, or −0, 100% tolerance.
±3, 6, 12 1/2, and 30% are ASA 40, 20, 10, and 5 step tolerances.

### Extended range T.C. tubular ceramics
Tolerance
1st, 2nd: significant figures
Multiplier

### Body-end band system
1st, 2nd: significant figures
Tolerance  Multiplier

# Capacitor Color Code

## Disc ceramics (5-dot system)
- 1st, 2nd significant figures
- Multiplier
- Tolerance
- Temperature coefficient

## Disc ceramics (3-dot system)
- 1st, 2nd significant figures
- Multiplier

### Ceramic capacitor codes (capacity given in MMF)

| Color | Digit | Multiplier | Tolerance 10 MMF or less | Tolerance Over 10 MMF | Temperature coefficient PPM/°C | Extended range Temp. Significant figure | Extended range Coeff Multiplier |
|---|---|---|---|---|---|---|---|
| Black | 0 | 1 | ±2.0 MMF | ±20% | 0(NP0) | 0.9 | -1 |
| Brown | 1 | 10 | ±0.1 MMF | ±1% | -33(N033) | | -10 |
| Red | 2 | 100 | | ±2% | -75(N075) | 1.0 | -100 |
| Orange | 3 | 1000 | | ±2.5% | -150(N150) | 1.5 | -1000 |
| Yellow | 4 | 10000 | | | -220(N220) | 2.2 | -10000 |
| Green | 5 | | ±0.5 MMF | ±5% | -330(N330) | 3.3 | +1 |
| Blue | 6 | | | | -470(N470) | 4.7 | +10 |
| Violet | 7 | | | | -750(N750) | 7.5 | +100 |
| Gray | 8 | 0.01 | ±0.25 MMF | | -30(P030) | | +1000 |
| White | 9 | 0.1 | ±1.0 MMF | ±10% | General purpose bypass and coupling | | +10000 |
| Silver | | | | | | | |
| Gold | | | | | +100(P100) (Jan) | | |

Ceramic capacitor voltage ratings are standard 500 volts, for some manufacturers, 1000 volts for other manufacturers, unless otherwise specified.

## High capacitance tubular ceramics insulated or non-insulated
- 1st, 2nd significant figures
- Multiplier
- Tolerance
- Voltage (optional)

## Temperature compensating tubular ceramics
- 1st, 2nd significant figures
- Multiplier
- Tolerance
- Temperature coefficient

## Current standard JAN and EIA code
- White (EIA) Black (JAN)
- 1st, 2nd significant figures
- Multiplier
- Tolerance
- Class or characteristics

## Button silver mica
- 1st (when applicable) sig. fig.
- 2nd for 1st
- 3rd for 2nd
- Multiplier
- Tolerance
- Class

## Molded flat paper capacitors (commercial code)
- 1st, 2nd significant figures
- Voltage
- Multiplier
- Tolerance
- Black or brown body

## Molded flat paper capacitors (JAN code)
- Silver
- 1st, 2nd significant figures
- Multiplier
- Tolerance
- Characteristic

## Molded ceramics
Using standard resistor color-code
- 1st, 2nd significant figure
- Multiplier
- White band Distinguishes capacitor from resistor

## Button ceramics
- 1st, 2nd significant figures
- Multiplier
- Tolerance

Viewed from soldered surface

## Stand-off ceramics
- 1st, 2nd significant figures
- Multiplier
- Tolerance
- Temperature coefficient

## Feed-thru ceramics
- 1st, 2nd significant figures
- Multiplier
- Tolerance
- Temperature coefficient

appendix three

# DIODE POLARITY IDENTIFICATION

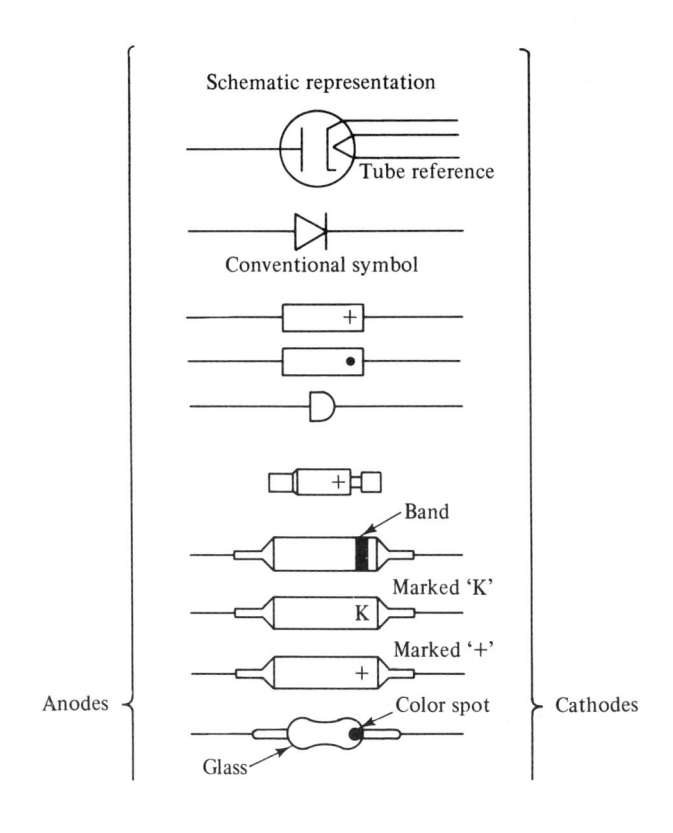

Schematic representation
Tube reference
Conventional symbol
Band
Marked 'K'
Marked '+'
Color spot
Glass
Anodes
Cathodes

# Diode Polarity Identification

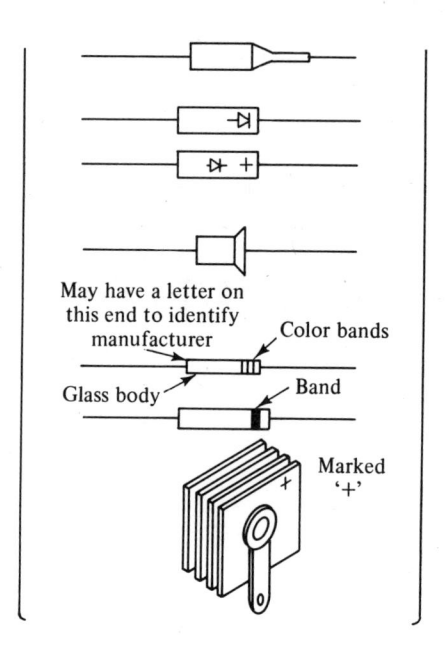

May have a letter on
this end to identify
manufacturer — Color bands

Glass body — Band

Marked
'+'

appendix four

# ELECTRONIC INDUSTRIES ASSOCIATION (EIA) PREFERRED VALUES FOR RESISTORS

Commercial tolerances on resistance values are 5, 10, and 20 percent. EIA preferred values are nominal resistance values that avoid duplication of stock within each tolerance range. The nominal resistance values tabulated below can be multiplied or divided by any power of 10.

**Tolerance**

| 5% | 10% | 20% |
|---|---|---|
| 10 | 10 | 10 |
| 11 | | |
| 12 | 12 | |
| 13 | | |
| 15 | 15 | 15 |
| 16 | | |
| 18 | 18 | |
| 20 | | |
| 22 | 22 | 22 |
| 24 | | |
| 27 | 27 | |
| 30 | | |
| 33 | 33 | 33 |
| 36 | | |
| 39 | 39 | 39 |
| 43 | | |
| 47 | 47 | 47 |
| 51 | | |

**Tolerance** (*Continued*)

| 5%  | 10% | 20% |
|-----|-----|-----|
| 56  | 56  |     |
| 62  |     |     |
| 68  | 68  | 68  |
| 75  |     |     |
| 82  | 82  |     |
| 91  |     |     |
| 100 | 100 | 100 |

appendix five

# EIA PREFERRED VALUES FOR ELECTROLYTIC CAPACITORS

±5% **Tolerance**

| Lowest Cap. | Nominal Preferred Value | Highest Cap. |
|---|---|---|
| * .95 | 1.0 | 1.05* |
| *1.045 | 1.1 | 1.155* |
| *1.14 | 1.2 | 1.26* |
| *1.235 | 1.3 | 1.365* |
| *1.425 | 1.5 | 1.575* |
| *1.52 | 1.6 | 1.68* |
| *1.71 | 1.8 | 1.89 |
| 1.90 | 2.0 | 2.1* |
| *2.09 | 2.2 | 2.31* |
| *2.28 | 2.4 | 2.52 |
| 2.565 | 2.7 | 2.835 |
| 2.85 | 3.0 | 3.15* |
| *3.135 | 3.3 | 3.465* |
| *3.42 | 3.6 | 3.78* |
| *3.705 | 3.9 | 4.095* |
| *4.085 | 4.3 | 4.515* |
| *4.465 | 4.7 | 4.935* |
| *4.845 | 5.1 | 5.355* |
| *5.320 | 5.6 | 5.880 |
| 5.890 | 6.2 | 6.510* |
| *6.460 | 6.8 | 7.140* |
| *7.125 | 7.5 | 7.875* |
| *7.790 | 8.2 | 8.610 |
| 8.645 | 9.1 | 9.555* |

# EIA Preferred Values for Electrolytic Capacitors

## ±10% Tolerance

| Lowest Cap. | Nominal Preferred Value | Highest Cap. |
|---|---|---|
| * .9 | 1.0 | 1.1* |
| *1.08 | 1.2 | 1.32* |
| *1.35 | 1.5 | 1.65* |
| *1.62 | 1.8 | 1.98* |
| *1.98 | 2.2 | 2.42 |
| 2.43 | 2.7 | 2.97* |
| *2.97 | 3.3 | 3.63* |
| *3.51 | 3.9 | 4.29* |
| *4.23 | 4.7 | 5.17* |
| *5.04 | 5.6 | 6.16* |
| *6.12 | 6.8 | 7.48* |
| *7.38 | 8.2 | 9.02* |

## ±20% Tolerance

| Lowest Cap. | Nominal Preferred Value | Highest Cap. |
|---|---|---|
| * .8 | 1.0 | 1.2* |
| *1.2 | 1.5 | 1.8* |
| *1.76 | 2.2 | 2.64* |
| *2.64 | 3.3 | 3.96* |
| *3.76 | 4.7 | 5.64* |
| *5.44 | 6.8 | 8.16* |

*Indicates overlap of tolerance spreads; highest capacitance of lower nominal preferred value equals or exceeds lowest capacitance of higher nominal value.

appendix six

# POWER RATIOS, VOLTAGE RATIOS,* AND DECIBEL VALUES

| Power Ratio | Voltage Ratio | Decibels −  + ←  → | Voltage Ratio | Power Ratio |
|---|---|---|---|---|
| 1.000 | 1.0000 | 0 | 1.000 | 1.000 |
| 0.9772 | 0.9886 | 0.1 | 1.012 | 1.023 |
| 0.9550 | 0.9772 | 0.2 | 1.023 | 1.047 |
| 0.9333 | 0.9661 | 0.3 | 1.035 | 1.072 |
| 0.9120 | 0.9550 | 0.4 | 1.047 | 1.096 |
| 0.8913 | 0.9441 | 0.5 | 1.059 | 1.122 |
| 0.8710 | 0.9333 | 0.6 | 1.072 | 1.148 |
| 0.8511 | 0.9226 | 0.7 | 1.084 | 1.175 |
| 0.8318 | 0.9120 | 0.8 | 1.096 | 1.202 |
| 0.8128 | 0.9016 | 0.9 | 1.109 | 1.230 |
| 0.7943 | 0.8913 | 1.0 | 1.122 | 1.259 |
| 0.6310 | 0.7943 | 2.0 | 1.259 | 1.585 |
| 0.5012 | 0.7079 | 3.0 | 1.413 | 1.995 |
| 0.3981 | 0.6310 | 4.0 | 1.585 | 2.512 |
| 0.3162 | 0.5623 | 5.0 | 1.778 | 3.162 |
| 0.2512 | 0.5012 | 6.0 | 1.995 | 3.981 |
| 0.1995 | 0.4467 | 7.0 | 2.239 | 5.012 |
| 0.1585 | 0.3981 | 8.0 | 2.512 | 6.310 |
| 0.1259 | 0.3548 | 9.0 | 2.818 | 7.943 |
| 0.10000 | 0.3162 | 10.0 | 3.162 | 10.00 |
| 0.07943 | 0.2818 | 11.0 | 3.548 | 12.59 |
| 0.06310 | 0.2512 | 12.0 | 3.981 | 15.85 |
| 0.05012 | 0.2293 | 13.0 | 4.467 | 19.95 |
| 0.03981 | 0.1995 | 14.0 | 5.012 | 25.12 |
| 0.03162 | 0.1778 | 15.0 | 5.623 | 31.62 |
| 0.02512 | 0.1585 | 16.0 | 6.310 | 39.81 |

*Voltage ratios based on equal load resistances.

| Power Ratio | Voltage Ratio | Decibels ← − | Decibels → + | Voltage Ratio | Power Ratio |
|---|---|---|---|---|---|
| 0.01995 | 0.1413 | 17.0 | | 7.079 | 50.12 |
| 0.01585 | 0.1259 | 18.0 | | 7.943 | 63.10 |
| 0.01259 | 0.1122 | 19.0 | | 8.913 | 79.43 |
| 0.01000 | 0.1000 | 20.0 | | 10.000 | 100.00 |
| $10^{-3}$ | $3.162 \times 10^{-2}$ | 30.0 | | $3.162 \times 10$ | $10^3$ |
| $10^{-4}$ | $10^{-2}$ | 40.0 | | $10^2$ | $10^4$ |
| $10^{-5}$ | $3.162 \times 10^{-3}$ | 50.0 | | $3.162 \times 10^2$ | $10^5$ |
| $10^{-6}$ | $10^{-3}$ | 60.0 | | $10^3$ | $10^6$ |
| $10^{-7}$ | $3.162 \times 10^{-4}$ | 70.0 | | $3.162 \times 10^3$ | $10^7$ |
| $10^{-8}$ | $10^{-4}$ | 80.0 | | $10^4$ | $10^8$ |
| $10^{-9}$ | $3.162 \times 10^{-5}$ | 90.0 | | $3.162 \times 10^4$ | $10^9$ |
| $10^{-10}$ | $10^{-5}$ | 100.0 | | $10^5$ | $10^{10}$ |

## appendix seven

# SYMBOLS FOR FIELD-EFFECT TRANSISTORS

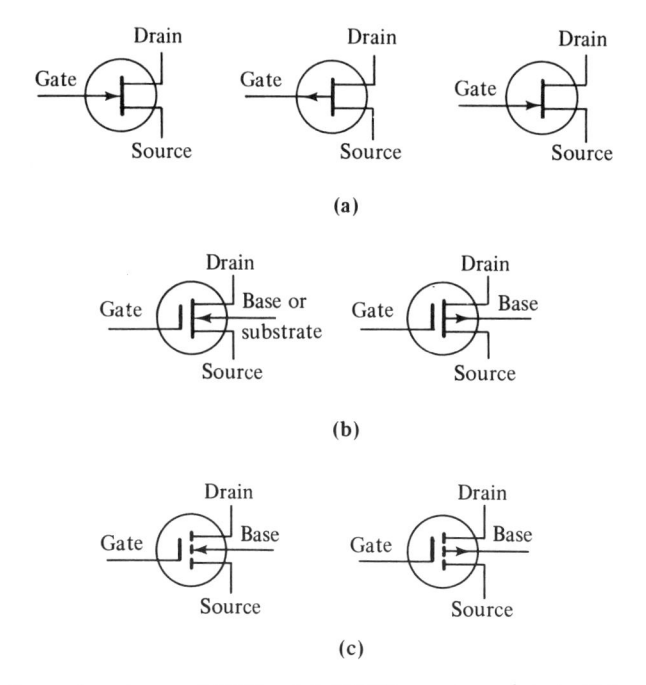

(a)

(b)

(c)

Symbols for various types of FET's: (a) JFET's: arrow points to N-type substance, away from P-type substance. (b) depletion-type MOSFET's: arrow points to N-type substrate, away from P-type substrate. (c) enhancement-type MOSFET's: arrow points to N-type substrate, away from P-type substrate.

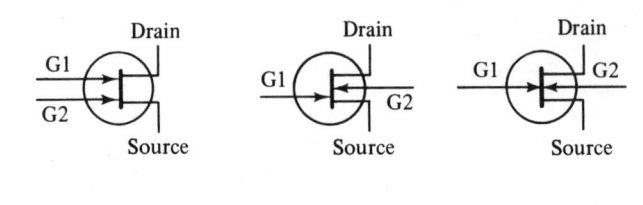

(d)

Symbols (*Continued*): (d) dual-gate FET's: N-channel symmetrical type is at left; N-channel nonsymmetrical type is at center; alternative symbol is at right.

## appendix eight

# RELATION OF DISTORTION TO POSITIVE FEEDBACK AND TO NEGATIVE FEEDBACK

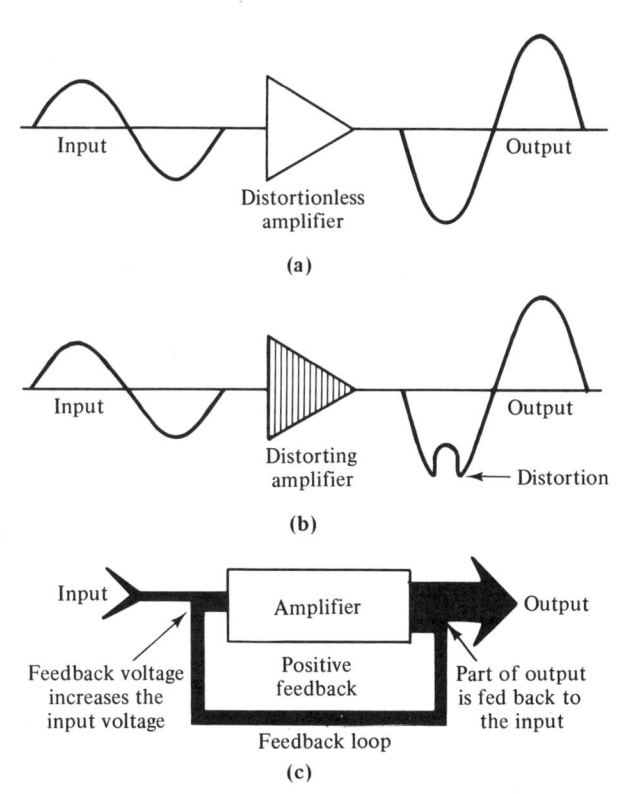

Positive feedback increases distortion: (a) amplifier with distortionless output; (b) amplifier with distorted output; (c) amplifier with positive-feedback loop.

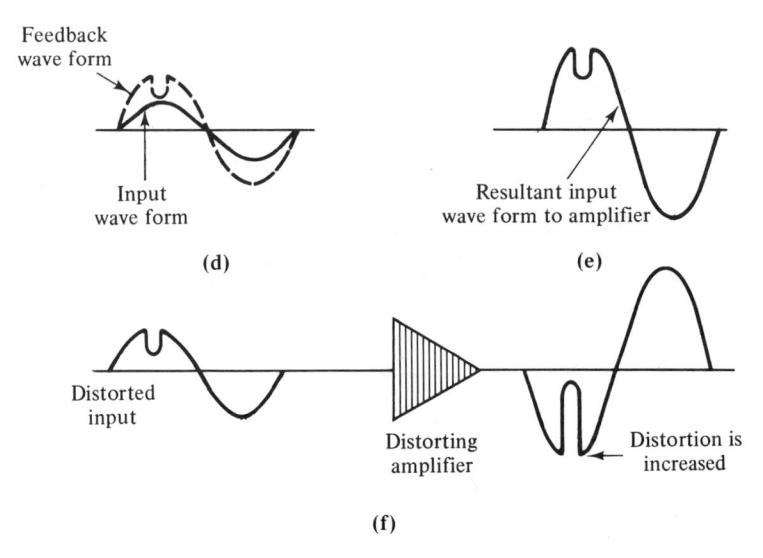

(d) input and feedback waveforms are in phase; (e) resultant input wave form to amplifier; (f) output waveform is further distorted by the amplifier.

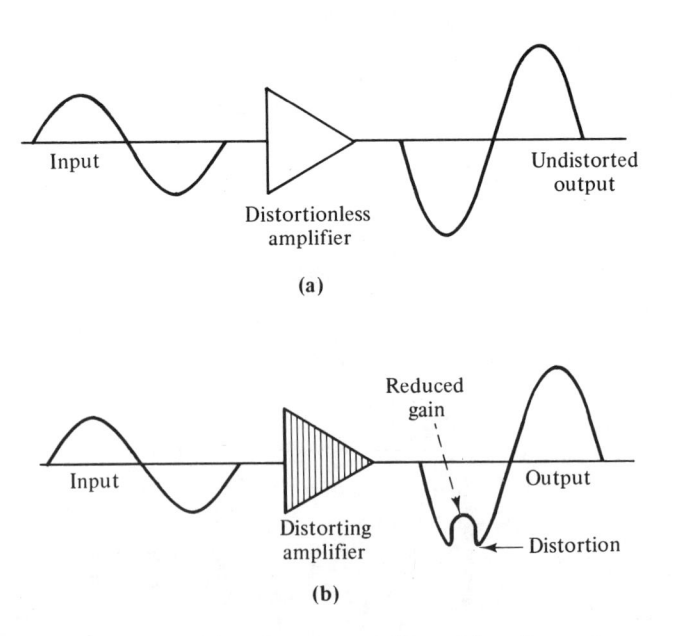

Negative feedback decreases distortion: (a) amplifier with undistorted output; (b) amplifier with distorted output.

## Relation of Distortion to Positive Feedback and to Negative Feedback

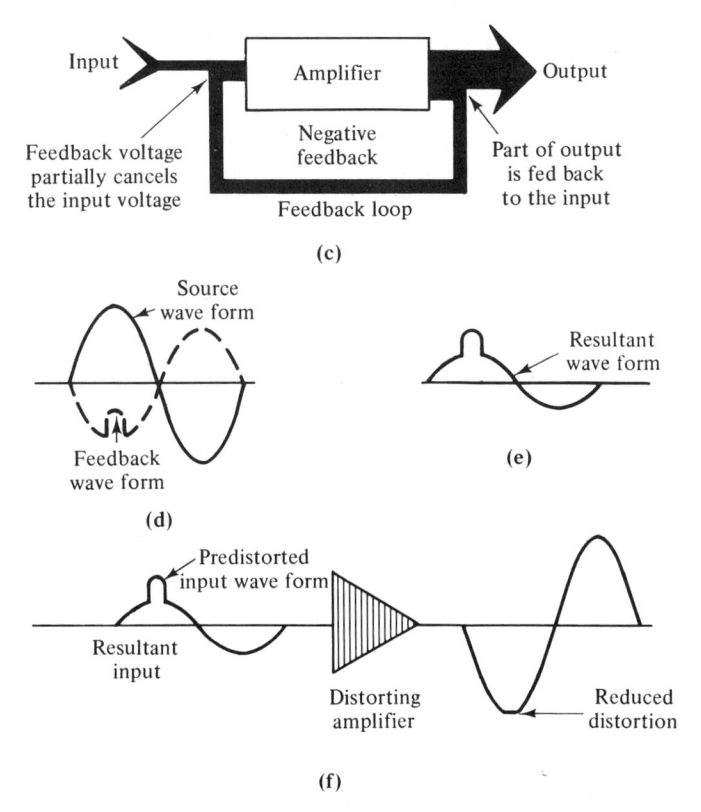

(c) amplifier with negative-feedback loop; (d) input and feedback waveforms are out of phase; (e) resultant input wave form to amplifier; (f) output waveform has reduced distortion.

# INDEX

## A

A, class, 23
AB, class, 23
Absolute maximum, 13
AC beta, 102
Acoustics, 209
  room, 188
Active
  devices, 39
  filter, 39
  region, 82
Agitation, thermal, 114
All-pass filter, 39
Alpha cutoff, 14
Ambient temperature, 16, 27, 132
Amplification, 210
  class A, 23
  class B, 23
  class D, 23
  class G, 23
  current, 72
  factor, 172
  power, 74
  voltage, 72

Amplifier, 210
  differential, 161
  direct coupled, 156
  low level, 130
  uncompensated, 36
  unity gain, 161
  unstabilized, 134
Amplitude distortion, 29
Analysis
  variational, 44
  worst case, 3
Angle, phase, 149
Antisidetone, 198
Audio, 210
  amplifier, 210
    classes, 23
    design, 65
  choke, 149
  circuit design, 39
  design phases, 2
  filter, 39
  frequency range, 42
  mixer, 42, 102

Audio (*Contd.*)
  room acoustics, 188
  voltmeter, 76
Autotransformer, 198

## B

B, class, 151
Bandpass, 57
  elimination filter, 39
  filter, 39
Bandwidth, power, 66
Base
  bias, 117
  current stabilization, 118
  resistance, 72
  voltage stabilization, 118
Beta, 70
  multiplier, 166
Bias
  circuitry, 121
  current, 133
  forward, 133
  requirements, 118
  reverse, 121
  shift, 88
  stability, 118
  stabilization, 117
    diode, 137
    resistor, 124
    thermistor, 133
    transistor, 141
  variation, 98
Black box, 29
Blocking capacitor, 159
Bogie
  cutoff frequency, 56
  value, 4, 9
Boost, low-frequency, 35
Bootstrapping, 114, 156
  loop, 114
Breakdown diode, 123, 146

Bridge
  circuit, 174
  transformer, 195
Bypass circuits, 42

## C

Calculated risk, 24
Capacitive
  load, 107
  reactance, 9, 197
Cascade operation, 104
Cascaded sections, 55, 116
CB, 68
CC, 68
CE, 68
Changeover distortion, 180
Characteristic impedance, 131
Charge carriers, 141
Circuit
  design, 39
  impedance, 9
Class
  A, 23
  AB, 23
  B, 151
  D, 177
  G, 180
Clipping distortion, 32, 91, 160
Collector resistance, 71
Common
  impedance, 25
  mode conduction, 166
Compensating circuit, 118
Complementary
  amplifier, 152
  drivers, 160
  symmetry, 74
Complete null, 60
Component tolerances, 14, 45
Cone offset, 145
Configuration taper, 20

# Index

Constant
  current source, 5, 68, 160
  phon value, 34
  voltage source, 5
    system, 191
Consumer electronics, 28
Corner frequency, 160
Coupling
  capacitance, 129
  circuits, 41
Cross
  connection, 141
  talk, 201
Crossover
  distortion, 32
  frequency, 184
  network, 184
Current
  amplification, 72
  stability factor, 125
Cutoff
  current, 14
  frequency, 45
  point, 90

## D

Darlington
  connection, 114
  pair, 166
DC beta, 102
D, class, 23
DC component, 42
Decade, 53
Decoupling
  circuit, 42
  network, 27
Degeneration, 104
Derating, 16
  factor, 98
Design
  center frequency, 46
  circuit, 39

Design (*Contd.*)
  innovation, 177
  phases, 1
  pitfalls, 25
  principles, 1
  reproducible, 1
Deviation, 47
Device
  isolation, 55
  tolerances, 14
Difference signal, 89
Differential amplifier, 161
Diode
  bias, 135
  characteristic, 81
  stabilization, 137
  zener, 146
Direct-coupled amplifier, 145, 156
Dissipation rating, 16
Distortion
  changeover, 180
  crossover, 32
  frequency, 29
  harmonic, 16
  intermodulation, 66
  parasitic, 29
  phase, 29
  total harmonic, 65
  transient, 166
Divider, voltage, 19
Double diode stabilization, 138
Drift, 161
Drive level, 77
Drop, voltage, 138
Dynamic
  range, 129
  resistance, 67

## E

Echoes, 204
Electron flow, 137
Electronics, consumer, 28

Efficiency, 70
Elimination filter, 39
Emitter
  bias, 121
  current, 131
  junction, 121
  region, 137
  resistance, 126, 129
  reverse bias, 131
  swamping resistor, 126
  terminal voltage, 131
Environment, 5
Environmental factors, 13
Equivalent circuit, 45
Expectancy, life, 5

**F**

Factor, gain, 58
Factors
  environmental, 13
  stability, 123
Feedback
  current, 104
  path, 26
  significant, 90
  tradeoff, 89
  voltage, 129
Filter, 216
  active, 39
  audio, 39
  bandpass, 39
  elimination, 39
  high-pass, 39
  low-pass, 39
  passive, 39
  rumble, 39
  scratch, 39
Flow
  bias current, 133
  electron, 137
  hole, 121
Forbidden regions, 13

Forward bias, 133
Four
  spiral, 201
  wire repeater, 204
Frequency, 217
  bogie cutoff, 56
  boost, 35
  compensating circuit, 42
  corner, 160
  crossover, 134
  cutoff, 45
  distortion, 29
  limiting, 77
  range, 42
  response, 9, 25
  zero, 139
Function, transfer, 88

**G**

Gain, 217
  current, 72
  factor, 58
  intrinsic, 9
  open-loop, 16
  power, 70
  reserve, 90
  unity, 161
  voltage, 16
Gaussian response, 116
G, class, 23
Generator resistance, 72
Graphical constructions, 80
Ground bus, 111

**H**

Harmonic, 218
  distortion, 16
  meter, 78
  total, 65
Hearing process, 22

# Index

Heat sink, 155
High
  fidelity, 30
  pass filter, 39
  power amplifier, 151
Holes, 121
  flow, 121
Hum, 218
Hybrid coil, 195

## I

Ideal bandpass, 57
Idling current, 155
IM, 66
Impedance, 9
  input, 9, 84
  output, 92
Induction coil, 195
Inductive load, 107
Infinite rolloff, 57
Innovative design, 177
Input
  and output relations, 21
  ports, 74
  resistance, 71
Insertion loss, 52
Instability, 25, 104
Integrated
  circuit, 16
  output, 178
Interaction, 42, 57
Intermodulation
  analyzer, 78
  distortion, 66
Internal
  impedance, 25, 107
  oscillation, 149
Intrinsic gain, 90
Iterated stages, 143

## J

Junction
  bias, 120
  diode, 144
  emitter, 121
  PN, 144
  resistance, 135
  saturation current, 121
  temperature, 98, 120
  transistor, 144

## L

Large-scale production, 1
Lattice structure, 135
Leakage current, 98, 156
Life-expectancy, 5
Limitations, stabilization, 67
Limiting
  frequencies, 77
  values, 81
Linear region, 82
Load, 219
  inductive, 107
  lines, 78
Loading principles, 202
Locus, 51
Loop
  bootstrap, 114
  feedback, 92
Loudness
  characteristic, 22
  control, 219
Low
  frequency boost, 35
    response, 11
  impedance microphone, 110
  level amplifier, 130
  pass filter, 39
  Z input, 74

## M

Majority carriers, 141
Manufacturers' ratings, 108
Manufacturing costs, 24
Matched pairs, 4, 74
Maximum
  deviation, 47
  efficiency, 70
  rated output, 75
Meter
  HD, 78
  IM, 78
Microphone, 110
Midband frequency, 104
Minority carriers, 126, 140
Mismatch, 169
Mixer, audio, 102
Mode, common, 166
Motorboating, 25
MRIA, 41
Multiple stage
  feedback, 107
  operation, 104
Multiple twin, 201
Music power, 67

## N

NAB, 40
Negative
  feedback, 16
  parallel-parallel, 92
  parallel-series, 92
  prototype arrangements, 93
  series-parallel, 92
  series-series, 92
Net
  impedance, 186
  resistance, 7
Network, 221
  crossover, 184
  decoupling, 27

Noise, 82
  figure, 14
  signals, 149
Nonlinear
  circuit operation, 79
  response, 35
  tapers, 20
Notch filter, 39, 59
Note, 221
Null, 59
  complete, 60
  shallow, 60
Nyquist plot, 106

## O

Occurrence, random, 22
Octave, 53, 221
One-tailed process, 24
Open-loop gain, 16
Operating
  environment, 5
  parameter, 107
  point, 81
    shift, 100
Operation
  cascade, 104
  multiple stage, 104
  small signal, 67
Oscillation, internal, 149
OTL, 152
Output
  capacitance, 14
  impedance, 92
  integrated, 178
  resistance, 71
  stabilization, 97
  stages, 32
  transformerless circuit, 152
Overcurrent, 28
Overdissipation, 155
Overload capability, 84, 87
Overtone, 221

# Index

## P

PA, 152
Parasitic distortion, 29
Passive filter, 39
Peak power, 112
Peaked response, 27
Percentage distortion, 76, 94
Performance specifications, 1
Phantom circuits, 201
Phase
   angle, 49
   characteristic, 105
   distortion, 29
   inverter, 152
   relation, 154
Phon units, 22, 34
Pitch, 34
Power
   amplification, 72, 74
   amplifier, 65
   bandwidth, 66, 75
   dissipation, 16
      capability, 20
   gain, 16, 70
   relations, 82
Preamplification, 74
Preamplifier, 18
Predistortion, 89
Probability, 23
   curve, 24
   distribution, 25
Production costs, 1
   tradeoffs, 109
Prototype
   arrangements, 92
   model, 3
Public address, 152
Push-pull, 151
PWM, 178

## Q

Quadded conductors, 201
Quadrature, 48
Quality control, 5
Quasi-complementary, 152
Quiescent point, 81
   shift, 100

## R

Random occurrence, 24
Range, dynamic, 129
Rated output, 75
RC circuits, 9
Reactance, 133
   capacitive, 197
   characteristic, 186
Reactive load, 45
Reconstituted signal, 151
Reflected voltage, 206
Regeneration, 27
Relation
   phase, 54
   power, 82
Reliability, 5
Repeaters, 202
Reproducible design, 1
Requirements, bias, 118
Reserve gain, 90
Resistance
   base, 72
   collector, 71
   dynamic, 67
   emitter, 129
   generator, 72
   junction, 135
   net, 7
   output, 71
Response
   frequency, 9, 25
   low frequency, 11

Response (*Contd.*)
  nonlinear, 35
  peaked, 27
Return loss, 205
Reverse bias, 121
  collector current, 118
RIAA, 41
Rolloff, 53
Room acoustics, 188
Rumble filter, 39

## S

Saturation current, 98, 113
  tolerance, 121
Scratch filter, 39
Self oscillation, 25, 105
Series-parallel feedback, 92
Series-series feedback, 92
Sidetone, 195
Shallow null, 60
Significant feedback, 90
Singing, 202
Small-signal operation, 67
Speaker, 225
  circuitry, 183
  interconnections, 184
  phasing, 190
Specifications, performance, 1
Spiral four, 201
Stability
  bias, 118
  factors, 123
Stabilization, 97
  temperature, 67
Stabilized configuration, 134
Stage feedback, 107
Standard configurations, 123
Stereophonic, 225
Strain relief, 28
Stretching distortion, 32
Surge protection, 149
Swamping resistor, 123

## T

Tapers, 20
Telephone circuitry, 195
Temperature
  coefficient, 13, 98
  stabilization, 67
Test procedures, 13
THD, 74
Thermal
  agitation, 14
  runaways, 121
Thermistor stabilization, 131
  limitations, 134
TIM, 33
Time constant, 9
Tolerance, 4
  deviations, 36
  requirements, 9
Tone control, 39, 61
  action, 39
Torsion, 12
Total harmonic distortion, 65
Transconductance, 16
Transfer
  characteristic, 21, 88, 99
  function, 88
Transformer
  coupling, 109
  distortion, 109
Transient
  distortion, 116
  intermodulation distortion, 33
  response, 156
Transistor
  deterioration, 108
  power dissipation, 108
T section, 7
Tweeter, 184
Two-tailed process, 24

# Index

## U

UL, 28
Uncompensated amplifier, 36
Underwriters' laboratories, 27
Unity gain amplifier, 161
Universal charts, 45
Unstabilized
 amplifier, 134
 circuit, 133
Utility sound, 191

## V

Variation
 collector current, 126
 emitter current, 125
Variational analysis, 5, 44
Voltage
 amplification, 72
 divider, 19
 drop, 16, 138

Voltage (*Contd.*)
 feedback, 129
 gain, 16
 regulator, 146
 stability, 146
 swing, 155

## W

Woofer, 184
Worst-case analysis, 3

## Y

Yield, 1

## Z

Zener diode, 146
Zero
 frequency, 39
 reference level, 34